Collins

KS3
Maths
Year 8

Workbook

Katherine Pate, Trevor Senior and
Michael White

About this Workbook

There are three Collins workbooks for KS3 Maths:
Year 7 Maths ISBN 9780008553692
Year 8 Maths ISBN 9780008553708
Year 9 Maths ISBN 9780008553715

Together they provide topic-based practice for all the skills and content on the Programme of Study for Key Stage 3 Maths.

The questions for each topic have been organised into **three levels** of increasing difficulty.

Track your progress by recording your marks in the box at the end of each level and the summary box at the end of each topic.

Found throughout the book, the **QR codes** can be scanned on your smartphone. Each QR code links to a video working through the solution to one of the questions on that double page spread.

Symbols are used to highlight questions that test key **skills**:

(MR) Mathematical Reasoning

(PS) Problem Solving

(FS) Financial Skills

To show how confident you feel with these skills, colour in the symbols alongside each question and at the end of each topic:

Green = Got it!
Orange = Nearly there
Red = Needs practice

Try to answer as many questions as possible without using a calculator. You will need extra paper for workings in some questions.

Questions where calculators **must not** be used are marked with this symbol:

The **answers** are included at the back so that you can mark your own work. **Helpful tips** are also included.

Contents

ACKNOWLEDGEMENTS

The authors and publisher are grateful to the copyright holders for permission to use quoted materials and images.

Every effort has been made to trace copyright holders and obtain their permission for the use of copyright material. The authors and publisher will gladly receive information enabling them to rectify any error or omission in subsequent editions. All facts are correct at time of going to press.

Published by Collins
An imprint of HarperCollins*Publishers* Ltd
1 London Bridge Street
London SE1 9GF

HarperCollins*Publishers*
Macken House
39/40 Mayor Street Upper
Dublin 1
D01 C9W8
Ireland

© HarperCollins*Publishers* Limited 2023
ISBN 9780008553708

First published 2023

10 9 8 7 6 5 4 3 2 1

Publisher: Clare Souza
Commissioning and project management: Richard Toms
Authors: Katherine Pate, Trevor Senior and Michael White
Cover Design: Sarah Duxbury and Kevin Robbins
Inside Concept Design: Sarah Duxbury and Paul Oates
Text Design and Layout: Contentra Technologies
Artwork: Contentra Technologies
Production: Emma Wood
Printed in India by Multivista Global Pvt. Ltd.

This book is produced from independently certified FSC™ paper to ensure responsible forest management.

For more information visit:
www.harpercollins.co.uk/green

Working with Numbers

1 Work out the value of: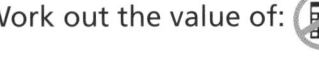

    a) 1^3 _____

    b) 8^2 _____

    c) $\sqrt{36}$ _____

    d) $\sqrt[3]{27}$ _____

    e) $\sqrt{16}$ _____

    f) 5^3 _____

    g) 2^4 _____

    h) $\sqrt[3]{8}$ _____ [8]

2 Work out:

    a) $2 \times 8 - 3 + 5$ _____

    b) $\dfrac{7+3}{5} - 2$ _____

    c) $4 + 3 \times 5$ _____

    d) $6^2 + 2 \times 5$ _____

    e) $8 + 5(4 + 7) - 6$ _____

    f) $2^3 (15 - 6) + 2 \times 3$ _____

    g) $\sqrt{49} + 2(3 + 5) - 4$ _____

    h) $\dfrac{40 + \sqrt{25}}{3^2}$ _____ [8]

3 Simplify each ratio.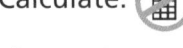

    a) $2 : 8 =$ _____

    b) $6 : 3 =$ _____

    c) $12 : 15 =$ _____

    d) $25 : 10 =$ _____

    e) $24 : 36 =$ _____

    f) $30 : 75 =$ _____ [6]

(FS) **4** Max buys 3 boxes of badges. Each box costs £10.50

There are 50 badges in each box. Max sells all the badges for 40p each.

How much profit does he make?

£ _____ [2]

(FS) **5** Lucy buys 5 apples, 4 bananas and a pack of strawberries.

Apples cost 24p each. Bananas cost 17p each.

Lucy pays with a £5 note and gets £1.02 change.

How much was the pack of strawberries?

£ _____ [2]

6 Calculate:

    a) -3×3 _____

    b) 8×-4 _____

    c) -5×-7 _____

    d) $18 \div -2$ _____

    e) $\dfrac{-24}{4}$ _____

    f) $-36 \div -6$ _____

    g) $(-5)^2$ _____

    h) $(-2)^3$ _____ [8]

7 Work out: 📵

    a) 4 – 9   b) –3 + 7

    c) –5 – 4   d) 5 – –3

    e) –6 + –2   f) –2 – –5

    g) 8 – 12 + 4   h) 17 + 3 – –6   [8]

Total Marks _____ / 42

1 Write all the prime numbers between 20 and 40

_____ [2]

2 Write each number as a product of prime factors.

    a) 56   b) 76   c) 88   d) 210

    _____ [4]

3 Find the HCF (highest common factor) and LCM (lowest common multiple) of:

    a) 14 and 20

                                               HCF: _____ LCM: _____

    b) 28 and 42

                                               HCF: _____ LCM: _____ [4]

(PS) 4 Here is a train timetable for Cardiff to London:

Cardiff	1047	1518	1855
London	1241	1712	2049

    a) How long (in hours and minutes) does the journey from Cardiff to London take?

                                                                   [1]

    b) Tom catches the 1855 from Cardiff. The train arrives in London 13 minutes late.

       At what time does it arrive in London?   [1]

Working with Numbers

(MR) **5** The Venn diagram shows the prime factors of 60 and 140

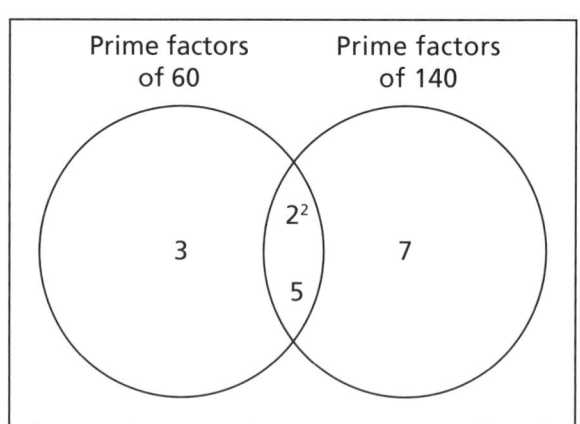

a) Find the HCF of 60 and 140

.... [1]

b) Find the LCM of 60 and 140

.... [1]

(PS) **6** Pink paint is made by mixing white paint and red paint in the ratio 4 : 1
Jack has 20 litres of white paint and 7 litres of red paint.

What is the maximum amount of pink paint he can make?

.... litres [2]

7 Sian swims one length of the pool every 25 seconds.
Toby swims one length of the pool every 40 seconds.
They both start at one end of the pool at the same time.

How long is it before they both reach an end of the pool at the same time?
Give your answer in minutes and seconds.

.... [2]

> **Total Marks** .... / 18

1 This table shows the flight distances, in kilometres, between cities in Europe.

Rome				
1353	**Madrid**			
1068	2846	**Kyiv**		
1530	2061	1320	**Copenhagen**	
1088	2380	1481	2142	**Athens**

What is the flight distance for each of these journeys?

a) Copenhagen to Madrid .... [1]

b) Rome to Kyiv .... [1]

c) Athens to Madrid and then Madrid to Rome .... [2]

2 Find the HCF and LCM of:

a) $2 \times 3^2 \times 5$ and $2^2 \times 3 \times 7$

HCF: .... [1]

LCM: .... [1]

b) 150 and 220

HCF: .... [1]

LCM: .... [1]

3 Work out: 🖩

a) $\sqrt[3]{-1}$ .... **b)** $\sqrt[3]{-1000}$ ....

c) $6 \times \sqrt[3]{125}$ .... **d)** $\dfrac{\sqrt[3]{-64}}{2}$ ....

e) $\dfrac{(3 + \sqrt{49})^2}{25}$ .... **f)** $\dfrac{12 - \sqrt[3]{8} - 20}{4 - 9}$ .... [6]

(FS) **4** Sophie makes and sells 6 kg of jam. To make 6 kg of jam, she uses 3 kg of fruit and 3 kg of sugar. She sells the jam in 500 g jars. Empty jam jars cost £1 for three.

1 kg of fruit costs £3.40 and 1 kg of sugar costs £0.78

She sells all her jars of jam for £2.30 per jar.

What is Sophie's total profit?

£ .... [5]

5 Alix and Dan share 400 g of sweets in the ratio 3 : 5

How many grams of sweets does each get?

Alix: .... g

Dan: .... g [2]

(FS) **6** Seb and Jo share some money in the ratio 5 : 1. Seb gets £16 more than Jo.

How much does Jo get?

£ .... [2]

Total Marks .... / 23

.... / 42

.... / 18

.... / 23

How do you feel about these skills?

(PS) (MR) (FS)

Green = Got it!
Orange = Nearly there
Red = Needs practice

Geometry

1 Shape P is shown on the coordinate grid.

a) Reflect shape P in the *y*-axis.
Label the image Q. [1]

b) Translate shape P to the
left by 3 units.
Label the image R. [1]

c) Reflect shape R in the *y*-axis.
Label the image S. [1]

d) Describe the translation of
shape Q to shape S.

.... [1]

(MR) **2** Two angles in a triangle are 30° and 75°. Alfie says that the triangle is isosceles.

Is he correct? Give a reason for your answer.

....

....

[2]

3

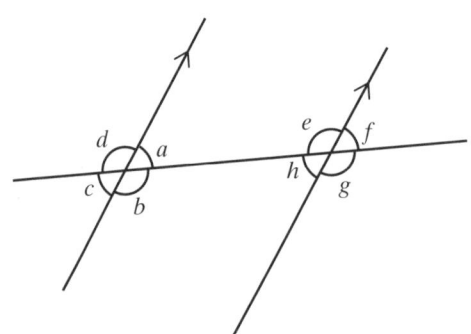

a) Which angle is **alternate** to angle *a*? .... [1]

b) Which angle is **corresponding** to angle *a*? .... [1]

c) Which angle is **vertically opposite** angle *e*? .... [1]

4 Which quadrilaterals below have diagonals of equal length?

(kite)   (rectangle)   (rhombus)   (square)

....

[2]

5 Find the sizes of the lettered angles.

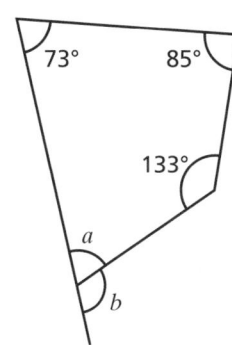

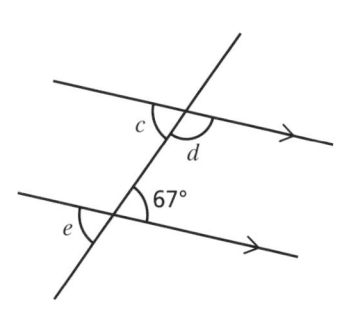

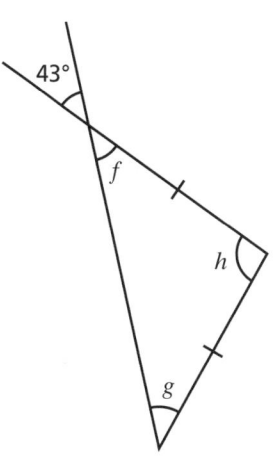

a = ... °   b = ... °   c = ... °   d = ... °

e = ... °   f = ... °   g = ... °   h = ... °   [8]

6 Triangle A is shown on the coordinate grid.

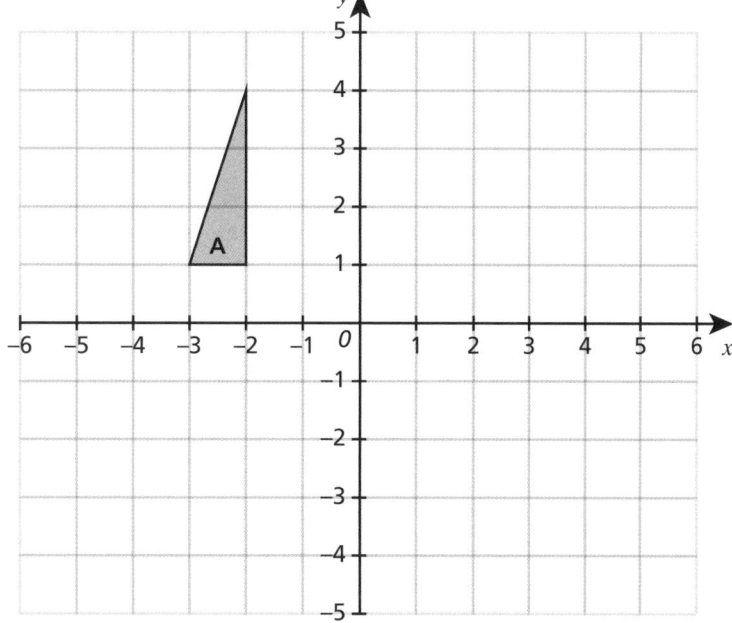

a) Reflect triangle A in the line x = 1. Label the image B.

Write down the coordinates of the vertices of B.   (........, ...), (........, ...), (........, ...) [2]

b) Reflect triangle B in the x-axis. Label the image C.

Write down the coordinates of the vertices of C.   (........, ...), (........, ...), (........, ...) [2]

c) Translate triangle C by 6 units to the left and 2 units up. Label the image D.

Write down the coordinates of the vertices of D.   (........, ...), (........, ...), (........, ...) [2]

d) Describe a rotation that takes triangle D back to the original triangle A.

.... [2]

Total Marks .... / 27

Geometry

 1 In quadrilateral ABCD, angle A is four times as large as angle D, angle B is three times as large as angle D and angle C is twice the size of angle D.

What is the size of angle D?

..................................° [3]

 2 A student constructs the perpendicular bisector of line AB as shown.

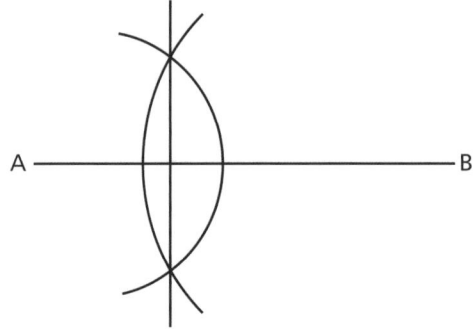

a) Explain what mistake the student has made.

....

.... [2]

b) What does the student's construction tell you about the diagonals of a kite?

.... [1]

 3 Use the diagram to explain why the angles in a quadrilateral sum to 360°.

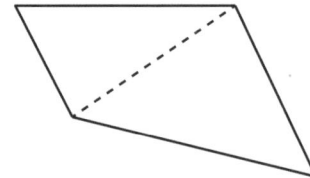

....

.... [2]

4 Use a ruler and compasses only to construct the angle bisector of the angle PQR. Show the construction lines clearly on the diagram.

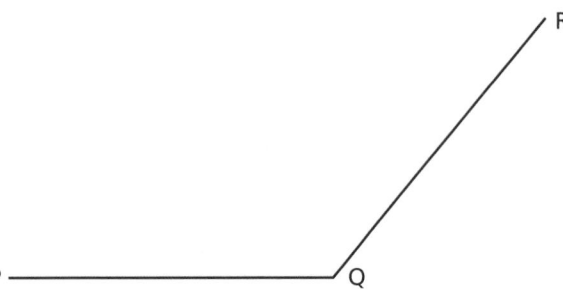

[2]

5 Work out the size of each lettered angle in the diagrams.

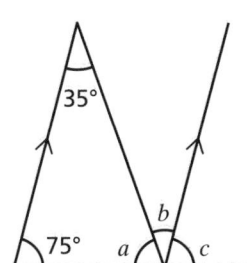
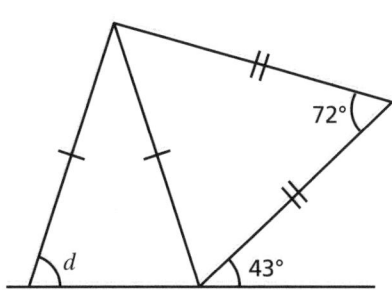
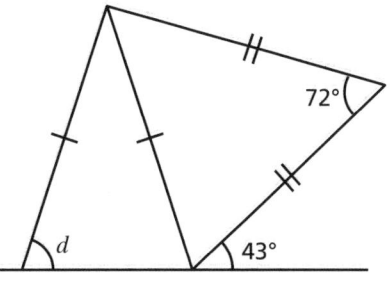

a = ...°   b = ...°   c = ...°   d = ...°   e = ...° [5]

6 Find the sizes of the lettered angles. Give a reason for each answer.

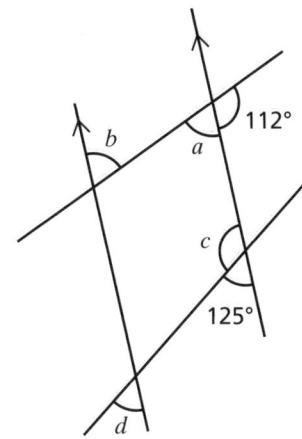

a = ...°

Reason: ....

b = ...°

Reason: ....

c = ...°

Reason: ....

d = ...°

Reason: ....

e = ...°

Reason: ....

f = ...°

Reason: ....

g = ...°

Reason: .... [14]

Geometry

(MR) **7** Explain why the lines PR and SU are **not** parallel.

....

....

....   [2]

(MR) **8** Prove that triangle ABD is isosceles.

....

....

....

....   [5]

Total Marks .... / 36

(MR) **1** A shape is rotated 180° about the origin (0, 0). Then the image is rotated 90° anti-clockwise about the origin. The new image is then rotated 180° about the origin to give the final image.

Describe the rotation that takes the final image back to the original shape.

....   [3]

2 Use a ruler and compasses only to construct a perpendicular line from point P to the line AB. Show the construction lines clearly on the diagram.

P

A ——————————————————————— B

[2]

3 In the diagram, BD is parallel to AF, and CE is parallel to BF.

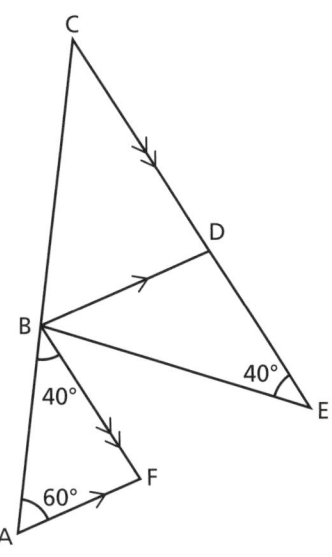

Prove that triangle BDE is isosceles.

....

....

....

....

.... [5]

4 The diagram shows a plan view of a school playground.

Ali starts at corner A and begins to walk across the playground. He walks along the straight line that is the angle bisector of AB and AD. He continues along this path until it intersects the perpendicular bisector of CD. He then walks along this perpendicular bisector until he reaches the mid-point of CD.

Draw Ali's route on the diagram, showing all construction lines.

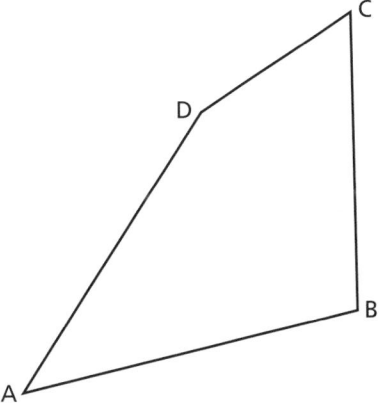

[5]

Geometry

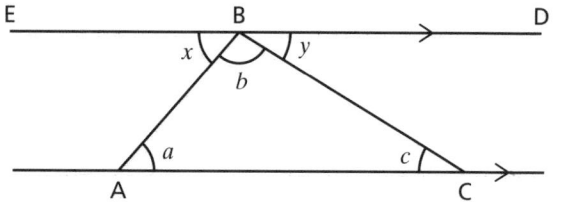
(MR) **5** EBD is a straight line and is parallel to AC.

Complete the proof. The first line is done for you.

$x = a$ Reason: Alternate angles are equal.

$y =$ ....   Reason: ....

$x + b + y =$ .... $^\circ$ Reason: ....

So $a + b + c =$ .... $^\circ$

So angles in a triangle add up to .... $^\circ$ [6]

(MR) **6** A pair of compasses is held on P, then on Q, to draw the two arcs shown.

By considering the dotted quadrilateral, explain clearly why a line drawn from A to B would be the perpendicular bisector of PQ.

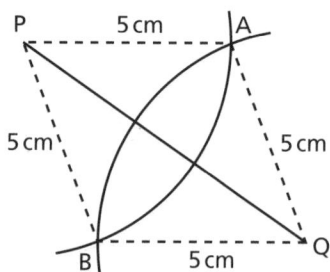

....

....

....

[3]

(MR) **7** Work out the size of angle PSR.

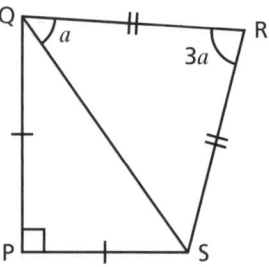

Angle PSR = ... $^\circ$ [4]

Total Marks .... / 28

.... / 27

.... / 36

.... / 28

Percentages

1 Work out:

a) 10% of 45 cm _____ cm   b) 50% of 500 ml _____ ml

c) 20% of £15   £ _____   d) 5% of 30 kg _____ kg

e) 25% of 160 m _____ m   f) 45% of 12 litres _____ litres   [6]

2 Write each fraction as a percentage.

a) $\frac{3}{25}$ = _____ %   b) $\frac{17}{20}$ = _____ %   c) $\frac{27}{40}$ = _____ %

d) $\frac{2}{5}$ = _____ %   e) $\frac{1}{8}$ = _____ %   f) $\frac{5}{8}$ = _____ %   [6]

3 Write each percentage as a decimal.

a) 15% = _____   b) 26% = _____   c) 120% = _____

d) 2% = _____   e) 8.3% = _____   f) 107% = _____   [6]

(MR) **4** Draw lines to match each question to the correct calculation.

Questions	Calculations
Work out 30% of £36	0.024 × 36
Work out 24% of £36	0.95 × 36
Work out 95% of £36	1.1 × 36
Work out 3% of £36	0.03 × 36
Work out 2.4% of £36	0.24 × 36
Work out 110% of £36	0.3 × 36

[4]

5 a) Increase £56 by 10%   £ _____   b) Increase 28 g by 30% _____ g

c) Increase 450 ml by 50% _____ ml   d) Increase 46 m by 5% _____ m   [4]

6 a) Decrease £25 by 10%   £ _____   b) Decrease 50 m by 25% _____ m

c) Decrease 32 kg by 40% _____ kg   d) Decrease 20 litres by 5% _____ litres   [4]

Total Marks _____ / 30

Percentages

Video Solution Question 6

1. Write each fraction as a percentage, to 1 decimal place.

a) $\frac{2}{3}$ = ... % b) $\frac{5}{7}$ = ... % c) $\frac{4}{9}$ = ... %

d) $\frac{25}{34}$ = ... % e) $\frac{49}{60}$ = ... % f) $\frac{173}{220}$ = ... % [6]

MR 2. Draw lines to match each question to the correct calculation.

Questions	Calculations
Increase £120 by 10%	1.4 × 120
Decrease £120 by 10%	0.96 × 120
Increase £120 by 4%	1.1 × 120
Decrease £120 by 4%	0.6 × 120
Increase £120 by 40%	1.04 × 120
Decrease £120 by 40%	0.9 × 120

[4]

3. In a class of 28 students:

- 15 walk to school
- 9 get the bus to school
- the rest cycle to school

What percentage, to the nearest whole number:

a) walk? .... % [2]

b) cycle? .... % [2]

FS 4. The value of a car increased from £5000 to £5600.

Work out the percentage increase in the value of the car.

.... % [1]

FS 5. The price of petrol decreased from £1.40 per litre to £1.25 per litre.

Work out the percentage decrease in the price of petrol.
Give your answer to 1 decimal place.

.... % [2]

Total Marks .... / 17

(PS) **1** A farmer has 200 chickens.
25% of the chickens are brown and $\frac{2}{5}$ of the chickens are white.
The rest of the chickens are black.

What percentage of the chickens are black? .... % [2]

(FS) **2** A phone costs £99.20 in a '20% off' sale.

What was the original price of the phone, before the sale?

£ .... [2]

(FS) **3** A builder charges £2430 to build a wall. 20% VAT is added to the cost.

What is the total cost?

£ .... [2]

(FS) **4** A meal in a restaurant costs £74.52 including 15% service charge.

How much does the meal cost excluding the service charge?

£ .... [2]

(FS) **5** In 2021 to 2022 house prices increased by 8%. In 2022, Toby's house was worth £240 000.

a) What was Toby's house worth in 2021, to the nearest pound?

£ .... [2]

b) Toby bought his house in 2015 for £195 000.

What was the percentage change in the price of his house from 2015 to 2022?
Give your answer to 1 decimal place.

.... % [2]

(MR) **6** Sam's cat is on a diet. After 2 months, the cat's mass decreases by 10%. It now weighs 6 kg.

Sam says that the cat must have weighed 6.6 kg before the decrease.
Explain why Sam is **not** correct.

....

.... [1]

Total Marks .... / 13

	/ 30
	/ 17
	/ 13

How do you feel about these skills?

(PS) (MR) (FS) Green = Got it!
Orange = Nearly there
Red = Needs practice

Sequences

1 Here is a number machine.

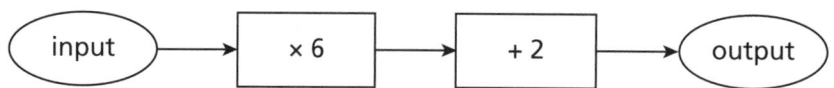

input → ×6 → +2 → output

a) Work out the output when the input is $\frac{1}{2}$

.... [1]

b) Work out the output when the input is –3

.... [1]

c) Work out the input when the output is 2

.... [1]

d) Work out the input when the output is –46

.... [1]

(MR) **2** Find the term-to term rule for each sequence.

a) 1, 5, 9, 13, 17, …   .... [1]

b) 1, 5, 25, 125, 625, …   .... [1]

3 Describe how each sequence is generated.

a) 1, 5, 10, 16, 21, …   .... [1]

b) 1, 5, 6, 11, 17, 28, …   .... [1]

4 Which of these sequences are arithmetic? Circle your answers.

A 4, 8, 12, 16, …

B 3, 9, 27, 81, …

C 19, 18, 17, 16, …

D 4, 2, 0, –2, …   [2]

5 The nth term of a sequence is $5n + 3$

a) Work out the value of the 10th term.   .... [1]

b) Decide which of these values are terms in the sequence.

125   88   103   206   .... [2]

(PS) **6** Here is a number machine.

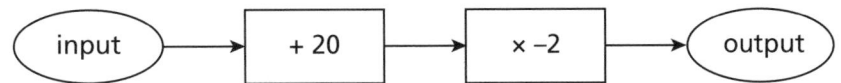

a) Use the number machine with inputs 1, 2, 3, 4 to create the first four terms
of a sequence.

.... [2]

b) Which term of the sequence is –60?

.... [2]

7 Write down the 1st, 2nd, 3rd, 10th, 50th and 100th terms of the sequences with nth terms
given by:

a) $3n + 4$

.... [3]

b) $6 - n$

.... [3]

c) $\frac{1}{2}n - 8$

.... [3]

d) $15 - 2n$

.... [3]

Total Marks .... / 29

1 Here are the nth terms of four arithmetic sequences.

Decide whether each sequence is increasing or decreasing.

a) $2n + 1$ .... [1]

b) $3n - 100$ .... [1]

c) $10 - n$ .... [1]

d) $-4n$ .... [1]

Sequences

(PS) **2** Here are some arithmetic sequences. In each part, work out the value of the letters.

a) 17, 20, a, b, …

$a =$ .... , $b =$ .... [2]

b) 19, c, 31, d, …

$c =$ .... , $d =$ .... [2]

c) 10, e, f, 4, …

$e =$ .... , $f =$ .... [2]

d) g, 9, h, 15, …

$g =$ .... , $h =$ .... [2]

3 Work out the nth term of the following sequences.

a) 4, 8, 12, 16, 20, …

.... [2]

b) 5, 9, 13, 17, 21, …

.... [2]

c) 18, 21, 24, 27, 30, …

.... [2]

d) 20, 18, 16, 14, 12, …

.... [2]

e) 3, 3.5, 4, 4.5, 5, …

.... [2]

f) 10, 9, 8, 7, 6, …

.... [2]

(MR) **4** The nth term of an arithmetic sequence is $5n + 3$

The nth term of a different arithmetic sequence is $7n - 5$

a) By listing terms of each sequence, show that 23 is in both sequences.

....

.... [2]

b) Find another number that is in both sequences.

.... [2]

(FS) **5** Jorge is saving money. He saves £2 the first week, £3 the second week, £4 the third week, and so on.

a) How much does he save in the 10th week?

£ .... [1]

b) How much does he save in the nth week?

£ .... [1]

c) He works out this formula for the total amount (A) he has saved after n weeks.
$A = \frac{n(n+3)}{2}$

Work out the total amount he has saved after 10 weeks.

£ .... [2]

d) Show that if he continued to save, the total saved after 19 weeks is over £200.

....

.... [2]

Total Marks .... / 34

(MR) **1** Here are two sequences.   99, 94, 89, 84, ... and 1, 4, 9, 16, ...

Which numbers will appear in both sequences? 🔲

.... [2]

Sequences

2 Here is a pattern of squares and circles.

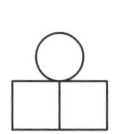 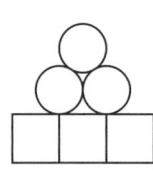 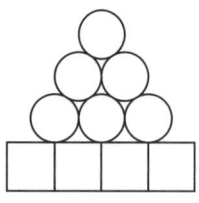 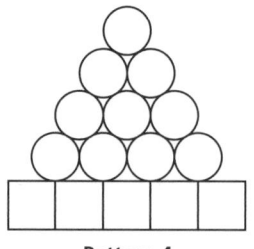

Pattern 1   Pattern 2   Pattern 3   Pattern 4

a) Complete the table to show the number of squares and circles in each pattern.   [2]

Pattern number	1	2	3	4
Squares				
Circles				

b) How many squares are in pattern n? .... [1]

c) What is the name given to the sequence for the number of circles?

.... [1]

d) A formula for the total number of circles (T) in pattern n is $T = \dfrac{n(n+1)}{2}$

Work out the number of circles in pattern 10.

.... [2]

e) Write down a formula for the total number of squares and circles (A)
in pattern n. You do not need to simplify it.

.... [1]

(PS) **3** Here is a sequence of coordinates. (1, 1), (4, 8), (9, 27), (16, 64), …

a) Work out the next coordinates in the sequence.

(............... , ...) [2]

b) Work out the nth coordinates in the sequence.

(............... , ...) [2]

Total Marks .... / 13

/ 29

/ 34

/ 13

How do you feel about these skills?

(PS) (MR) (FS)

Green = Got it!
Orange = Nearly there
Red = Needs practice

Area and Volume

1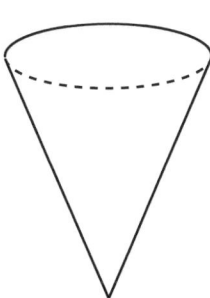

a) Explain why this solid is **not** a prism.

.... [1]

b) Write down the name of any solid that is
a prism. .... [1]

2 Work out the volume of this triangular prism.

Volume of a prism = Area of cross section × length of prism
$$V = Al$$

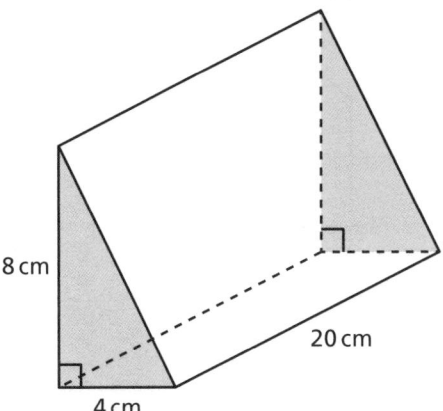

8 cm

20 cm

4 cm

.... cm³ [2]

(MR) **3** P is a 3D shape with uniform cross-section along its length. The volume of P is equal to the volume of cuboid C.

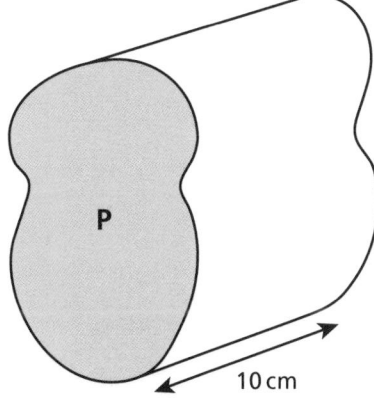

P

10 cm

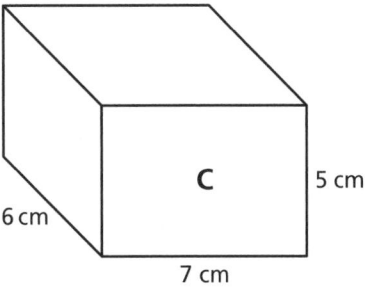

C   5 cm

6 cm

7 cm

Work out the area of the shaded cross-section of P.

.... cm² [4]

(PS) **4** Which shape below has the greater area and by how much?

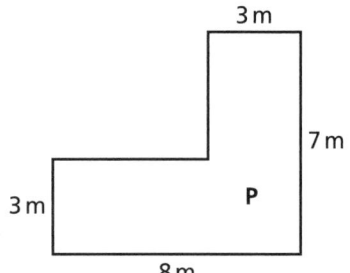

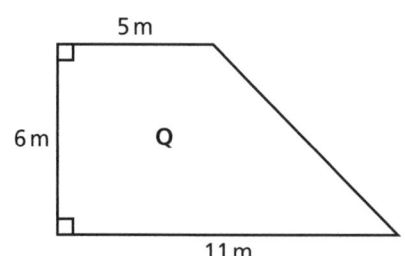

[3]

(PS) **5** Which cuboid below has the greater surface area and by how much?

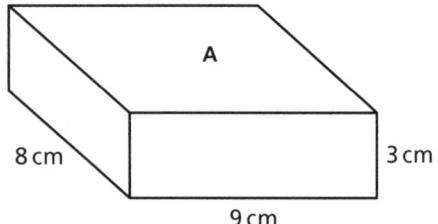

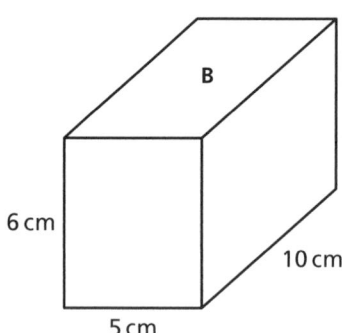

[5]

6 **a)** Sketch all the faces of this triangular prism. [2]

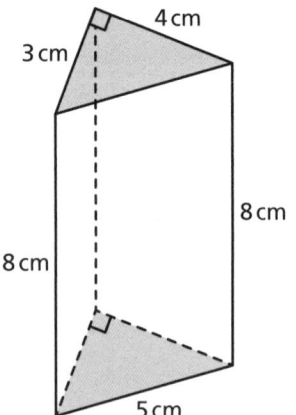

b) Work out the total surface area of the triangular prism.

.... cm² [3]

Total Marks .... / 21

1 Work out the volume of this prism.

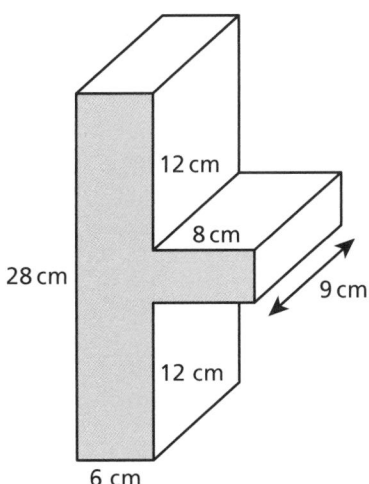

12 cm

8 cm

28 cm

9 cm

12 cm

6 cm

.... cm³ [4]

2 This cuboid is to be sprayed with varnish. The cost of spraying 40 cm² is £1.

Work out the cost of spraying the entire surface area.

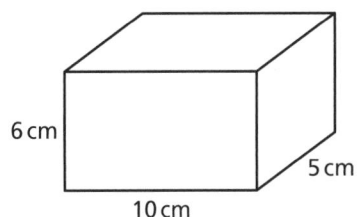

6 cm

5 cm

10 cm

£ .... [4]

3 Ava has a rectangular garden with a flower bed.
She wants to returf the lawn. Turf costs £6.75 per square metre.

How much will it cost Ava to returf her lawn?

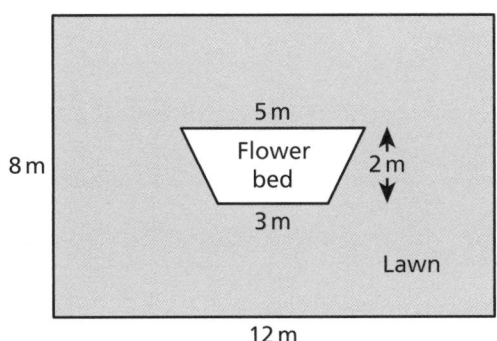

5 m

Flower
bed

2 m

8 m

3 m

Lawn

12 m

£ .... [4]

4 Work out the area of shape ABCDE.

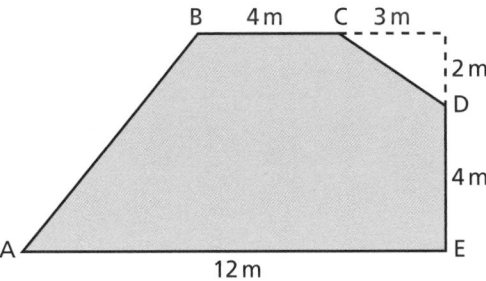

B   4 m   C   3 m

2 m

D

4 m

A

12 m

E

.... m² [3]

25

Area and Volume

(MR) **5** The volume of this triangular prism is 2160 cm³.

By finding the length, *l*, of the prism, work out its total surface area.

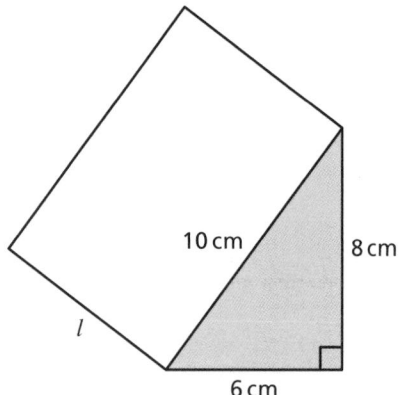

10 cm

8 cm

l

6 cm

.... cm² [6]

6 Work out the area of this trapezium.

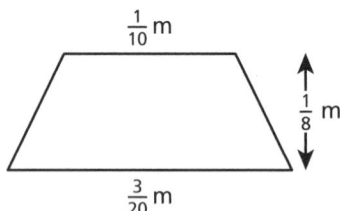

$\frac{1}{10}$ m

$\frac{1}{8}$ m

$\frac{3}{20}$ m

.... m² [2]

Total Marks .... / 23

(PS) **1** Work out the total surface area of this prism.

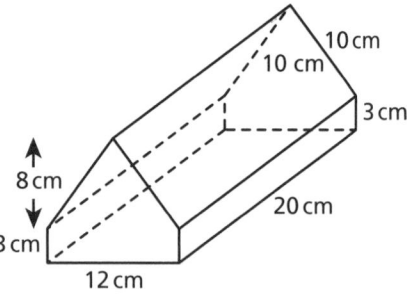

10 cm

10 cm

3 cm

8 cm

3 cm

20 cm

12 cm

.... cm² [4]

(MR) **2** A cube has a total surface area of 96 cm².

Work out the volume of the cube.

.... cm³ [4]

(MR) **3** The two rectangles shown are made on the ground with a thin layer of sand. The sand is swept up then it is all used to make one square.

Write down an expression, in terms of n, for the length of one side of the square.

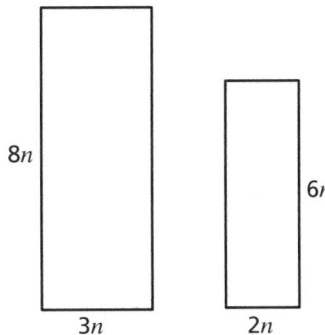

_____ [3]

(PS) **4** The diagram shows a 12 m square garden. The shaded triangle is to be a decking area.
(MR) The cost of the decking is £209.80 for every $4 \, m^2$.
(FS)

Work out the total cost of the decking area.

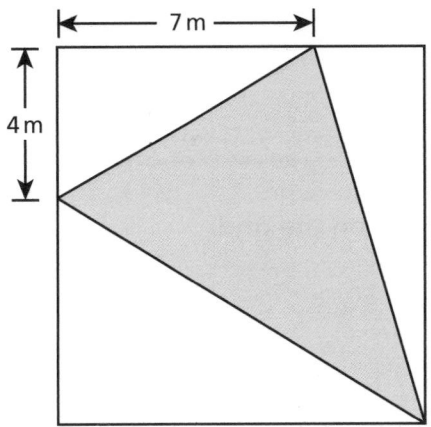

£ _____ [6]

Total Marks _____ / 17

_____ / 21

_____ / 23

_____ / 17

How do you feel about these skills?
(PS) (MR) (FS)   Green = Got it! Orange = Nearly there Red = Needs practice

Graphs

1 **a)** Complete the table of values for $y = x$

x	−2	−1	0	1	2
y					

[1]

b) Complete the table of values for $y = -x$

x	−2	−1	0	1	2
y					

[1]

c) On the grid, draw and label the graphs of $y = x$ and $y = -x$

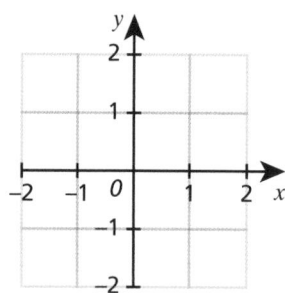

[2]

2 Write down the coordinates of the y-intercepts for each line on the grid.

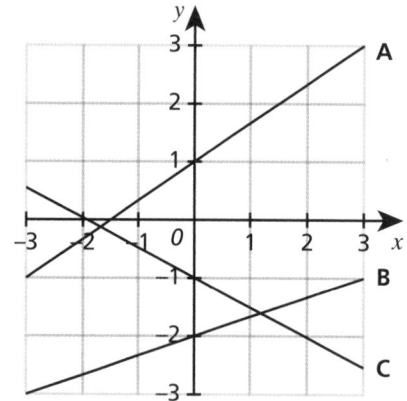

A (............ , ...) [1]

B (............ , ...) [1]

C (............ , ...) [1]

(MR) **3** **a)** Complete the table for the equations shown.

x	−2	−1	0	1	2
$y = 2x$	−4				
$y = 2x + 2$				4	
$y = 2x + 1$					5
$y = 2x - 1$		−3			
$y = 2x - 2$			−2		

[5]

b) Draw and label the graph for each equation in the table.

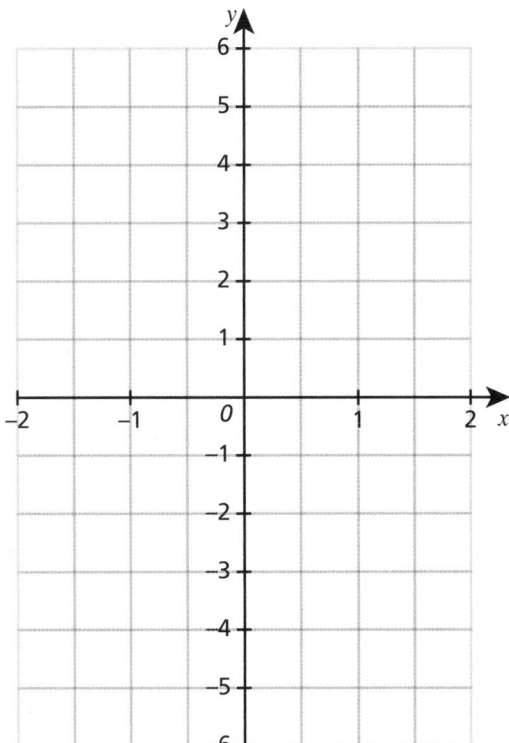

[5]

c) What property do you notice about the lines?

.... [1]

d) Write down the equation of a different line that has the same property as the lines above.

.... [1]

4 For each part, work out the gradient of the line. Parts a) and b) have been done for you.

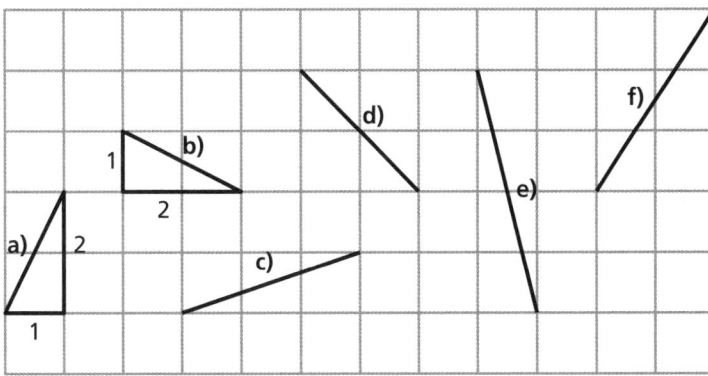

a) Gradient = $\frac{2}{1}$ = 2

b) Gradient = $-\frac{1}{2}$

c) Gradient = ...

d) Gradient = ...

e) Gradient = ...

f) Gradient = ... [4]

Total Marks .... / 23

Graphs

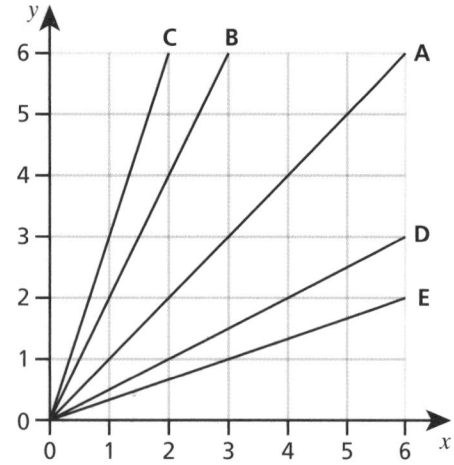

1 Work out the gradient for each line on the grid.

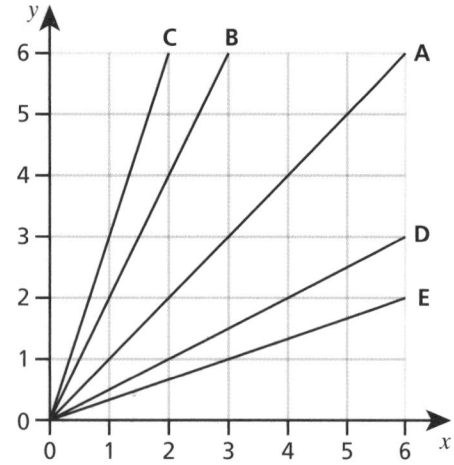

A Gradient = _____ [1]

B Gradient = _____ [1]

C Gradient = _____ [1]

D Gradient = _____ [1]

E Gradient = _____ [1]

2 Write down the gradient and the coordinates of y-intercepts of these lines. An example has been done for you.

Example: $y = 3x + 2$ Gradient = 3 y-intercept is (0, 2)

a) $y = 4x + 1$ Gradient = _____ y-intercept is (_____, _____) [1]

b) $y = 4x - 1$ Gradient = _____ y-intercept is (_____, _____) [1]

c) $y = 5x - 3$ Gradient = _____ y-intercept is (_____, _____) [1]

d) $y = \frac{1}{2}x + 2$ Gradient = _____ y-intercept is (_____, _____) [1]

e) $3x + y = 4$

Gradient = _____ y-intercept is (_____, _____) [2]

f) $2y + x = -1$

Gradient = _____ y-intercept is (_____, _____) [2]

3 Write down the equation of the straight line with:

a) a gradient of 3 passing through (0, 1) _____ [1]

b) a gradient of 3 passing through (0, −2) _____ [1]

c) a gradient of 4 passing through (0, −3) _____ [1]

d) a gradient of −2 passing through (0, 6) _____ [1]

e) a gradient of $\frac{1}{3}$ passing through (0, 4) _____ [1]

f) a gradient of $-\frac{1}{3}$ passing through (0, −2) _____ [1]

4 For each grid, write down the equations of lines A, B and C in the form $y = mx + c$.

a)
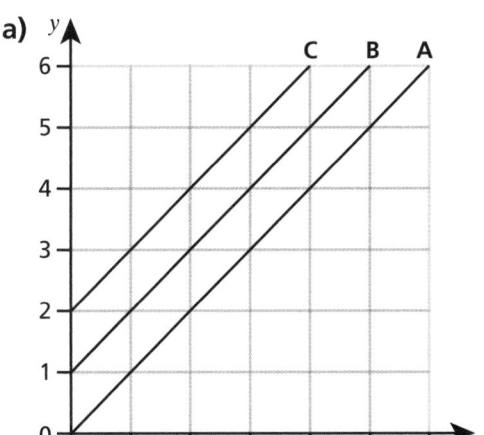

A .... [1]

B .... [1]

C .... [1]

b)
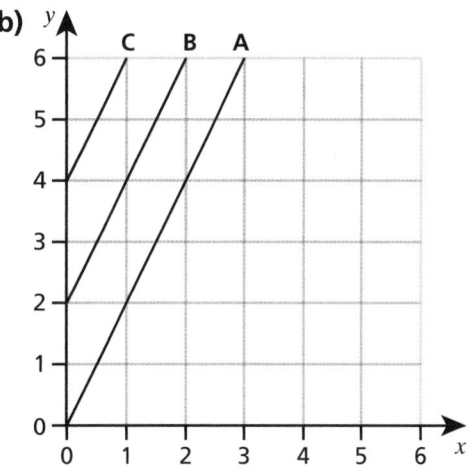

A .... [1]

B .... [1]

C .... [1]

c)
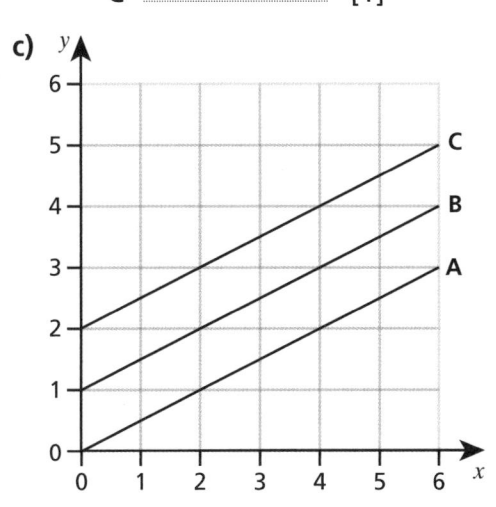

A .... [1]

B .... [1]

C .... [1]

Graphs

(MR) **1** Match each line to its correct equation.

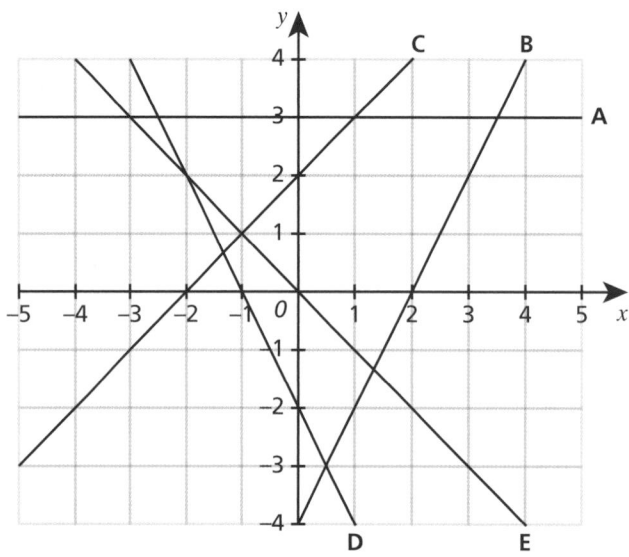

| y = 3 | | y = −x | | y = 2x − 4 | | y = −2x − 2 | | y = x + 2 |

[3]

(MR) **2** Here are the equations of eight straight lines.

y = 3x + 3   y = −3x + 2   y = 3x − 4   y = −3x − 4

y = −3x + 3   y = 2x − 3   y = 3x − 3   y = −3x − 3

a) Write each equation in its correct column in the table. The first one has been done for you.

Lines with gradient 3	Lines with gradient −3	Lines passing through (0, 3)	Lines passing through (0, −3)
y = 3x + 3		y = 3x + 3	

[4]

b) Write down the equation of a straight line that is parallel to $y = 3x + 3$
   and passes through (0, 7). .... [1]

Total Marks .... / 8

.... / 23

.... / 28

.... / 8

How do you feel about these skills?
(PS)   (MR)

Green = Got it!
Orange = Nearly there
Red = Needs practice

Simplifying Numbers

1 Round:

    a) 7.46 to 1 decimal place   [1]

    b) 8.034 to 2 decimal places   [1]

    c) 15.3825 to 3 decimal places   [1]

    d) 0.0316 to 2 decimal places   [1]

    e) 27.309 to 1 decimal place   [1]

    f) 6.497 to 2 decimal places   [1]

2 Round each number to 1 significant figure.

    a) 58   **b)** 127

    c) 4785   **d)** 60792   [4]

3 Round each number to 2 significant figures.

    a) 364   **b)** 4092

    c) 56231   **d)** 20800

    e) 352005   **f)** 207513

    g) 1496000   **h)** 13450000   [8]

4 Round each number to 3 significant figures.

    a) 6721   **b)** 98450

    c) 3159   **d)** 408612

    e) 35460   **f)** 105720

    g) 2413017   **h)** 1370946   [8]

5 Round each number to 1 significant figure and then multiply to estimate the answer to each calculation. 🖩

    a) 24 × 13   [1]

    b) 380 × 21   [1]

    c) 426 × 28   [1]

    d) 5706 × 32   [1]

Total Marks   / 30

Simplifying Numbers

1 Round each number to 1 significant figure.

a) 2.47 ....   b) 3.09 ....

c) 16.3 ....   d) 0.25 ....

e) 0.032 ....   f) 0.0056 ....

g) 0.00021 ....   h) 0.0508 .... [8]

2 Round each number to the given level of accuracy.

Number	1 s.f.	2 s.f.	3 s.f.
147			
2095			
18.36			
8079			
0.0351			
4872			
0.7599			

[7]

3 Round 3547.862 to:

a) 3 s.f. ....   b) the nearest tenth ....   c) 1 s.f. ....

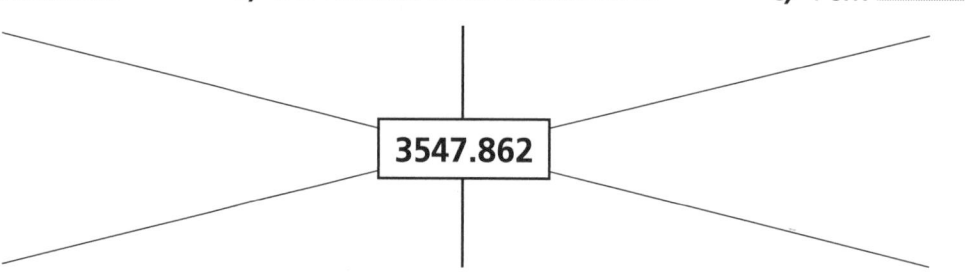

3547.862

d) the nearest 100 ....   e) 2 d.p. ....   f) 1 d.p. .... [6]

4 Estimate the answers to these calculations.

a) $\dfrac{4.76 \times 18}{9.7}$ .... [1]

b) $\dfrac{87.2 + 31.8}{42}$ .... [1]

c) $\dfrac{7.3}{0.51}$ .... [1]

d) $(0.19)^2$ .... [1]

Total Marks .... / 25

(MR) **1** **a)** Kay rounds 19 542 to 19 500

How many significant figures has she rounded it to? .... [1]

b) Josh rounds 19 542 to 2 significant figures.
Lee rounds 19 542 to 1 significant figure.

What do you notice about their answers?

....

.... [2]

2 Estimate the answers to these calculations.

a) $(27.5)^2 - 278$ .... [1]

b) $\dfrac{0.39 \times 7.8}{0.43}$ .... [1]

3 Estimate the area of this rectangle:

177 m

540 m

.... m^2 [1]

(PS) **4** **a)** Estimate the volume of a cube of side 4.8 cm.

(MR)
.... cm^3 [1]

b) Is your estimate an overestimate or an underestimate? .... [1]

(PS) **5** A box of chocolate bars weighs 18.45 kg.
One chocolate bar weighs 225 g.

Estimate the number of chocolate bars in the box.

.... [2]

Total Marks .... / 10

....	/ 30
....	/ 25
....	/ 10

How do you feel about these skills?
(PS) (MR) Green = Got it! Orange = Nearly there Red = Needs practice

Presenting and Interpreting Data

1 The table shows the times, to the nearest minute, that 20 people spent waiting for a bus.

Time (to the nearest minute)	0	1	2	3	4	5	6
Number of people	2	5	0	4	2	6	1

a) Complete the bar chart to represent this information.

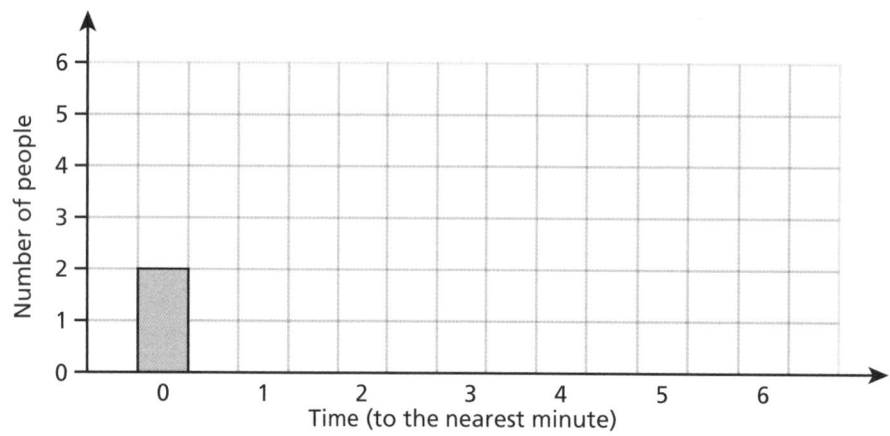

[2]

b) Work out the mean waiting time.

_____ minutes [3]

2 The pie chart shows the types of tickets used on a bus during one day.
Altogether 180 people used the bus.

Work out the number of each type of ticket used.

Number of tickets

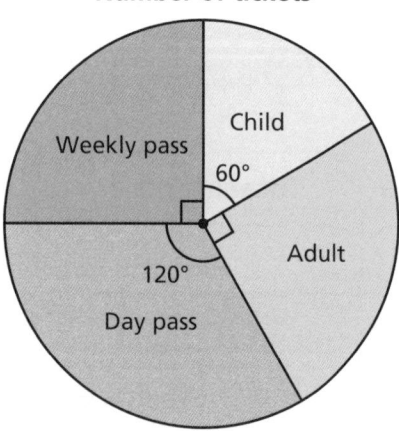

Child: _____

Adult: _____

Day pass: _____

Weekly pass: _____ [4]

3 The pie chart shows how 60 students travel to school.

Travel to school

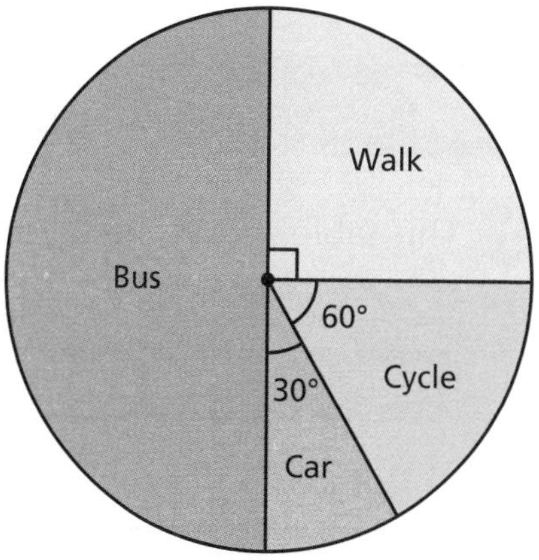

a) How many students get to school by bus?

.... [1]

b) Complete the pictogram to show how the students travel to school.

Bus	
Walk	
Cycle	
Car	

Key:

☐ represents 10 students

[3]

4 A group of Year 7 students were asked to choose their favourite animal from dog, cat, horse or llama.

Favourite animal

The pie chart shows the percentages chosen for each animal.

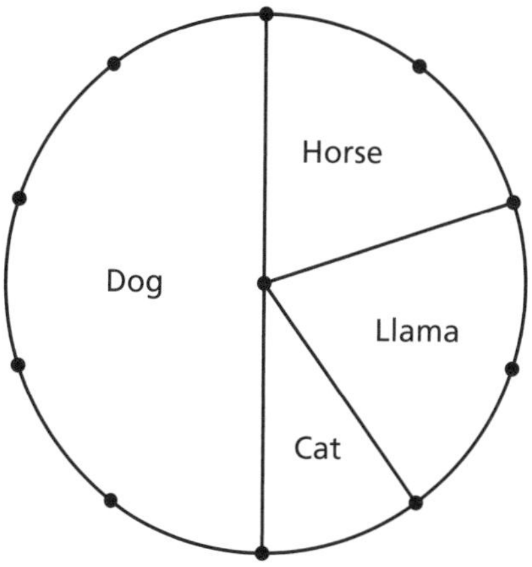

a) Work out the percentage of students that chose each animal.

Dog: .... %

Horse: .... %

Llama: .... %

Cat: .... % [4]

b) A group of Year 8 students were also asked and 40% chose dog, 25% chose cat and 5% chose horse.

What percentage of the Year 8 students chose 'llama'?

.... % [2]

c) Which of the following statements, comparing the Year 7 and Year 8 responses, are **definitely true**? For any that are not definitely true, give a reason for your answer.

i) More Year 7 students chose 'dog'.

.... [1]

ii) If the results are combined, 'dog' was the most popular choice.

.... [1]

iii) A greater proportion of Year 8 students chose 'cat'.

.... [1]

Total Marks .... / 22

Presenting and Interpreting Data

1) This table shows the favourite colours of 24 students.

Colour	Red	Blue	Yellow	Green
Frequency	6	12	5	1

a) Complete this table of angle sizes to show the information given above in a pie chart.

Number of students	24	1	5	6	12
Angle size	360°				

[2]

b) Draw a fully labelled pie chart.

Favourite colour

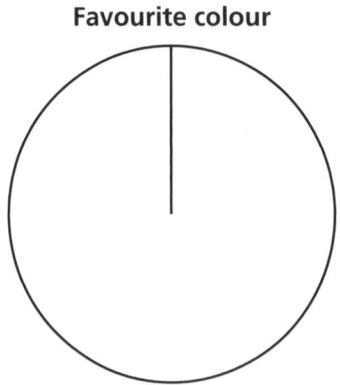

[2]

(MR) 2) The table shows the ways that some students travel to school.

	Walk or cycle	Bus or car
Year 7	140	40
Year 8	125	55
Year 9	99	81

a) How do the numbers change as students get older?

....

.... [1]

b) Between which two years is the greater change?
Show your working.

.... [1]

c) If the trend continues, approximately how many students do you expect to walk or cycle
in Year 10?

.... [2]

(PS) **3** The smallest sector on a pie chart is 45° and represents 15 students.

a) How many students are represented in the whole pie chart?

.... [2]

b) Tony says, "The biggest sector on the pie chart is 100°."

Is that possible? Show working to support your answer.

....

.... [2]

(MR) **4** A cyclist compares data about some of their rides.

For each scatter diagram, describe what each graph tells you.

a) Comparison of distance and time taken

What the graph tells you:

....

....

.... [1]

b) Comparison of distance and amount of rainfall

What the graph tells you:

....

....

.... [1]

c) Comparison of distance and number of people on ride

What the graph tells you:

....

....

.... [1]

Total Marks .... / 15

39

1 Here are the results for 15 students in two mathematics tests.

Student	A	B	C	D	E	F	G	H	I	J	K	L	M	N	P
Test X	20	50	75	15	45	80	27	32	21	64	35	40	56	34	68
Test Y	30	42	80	20	44	77	35	40	15	68	40	50	46	43	64

a) Draw a scatter graph to show these results.

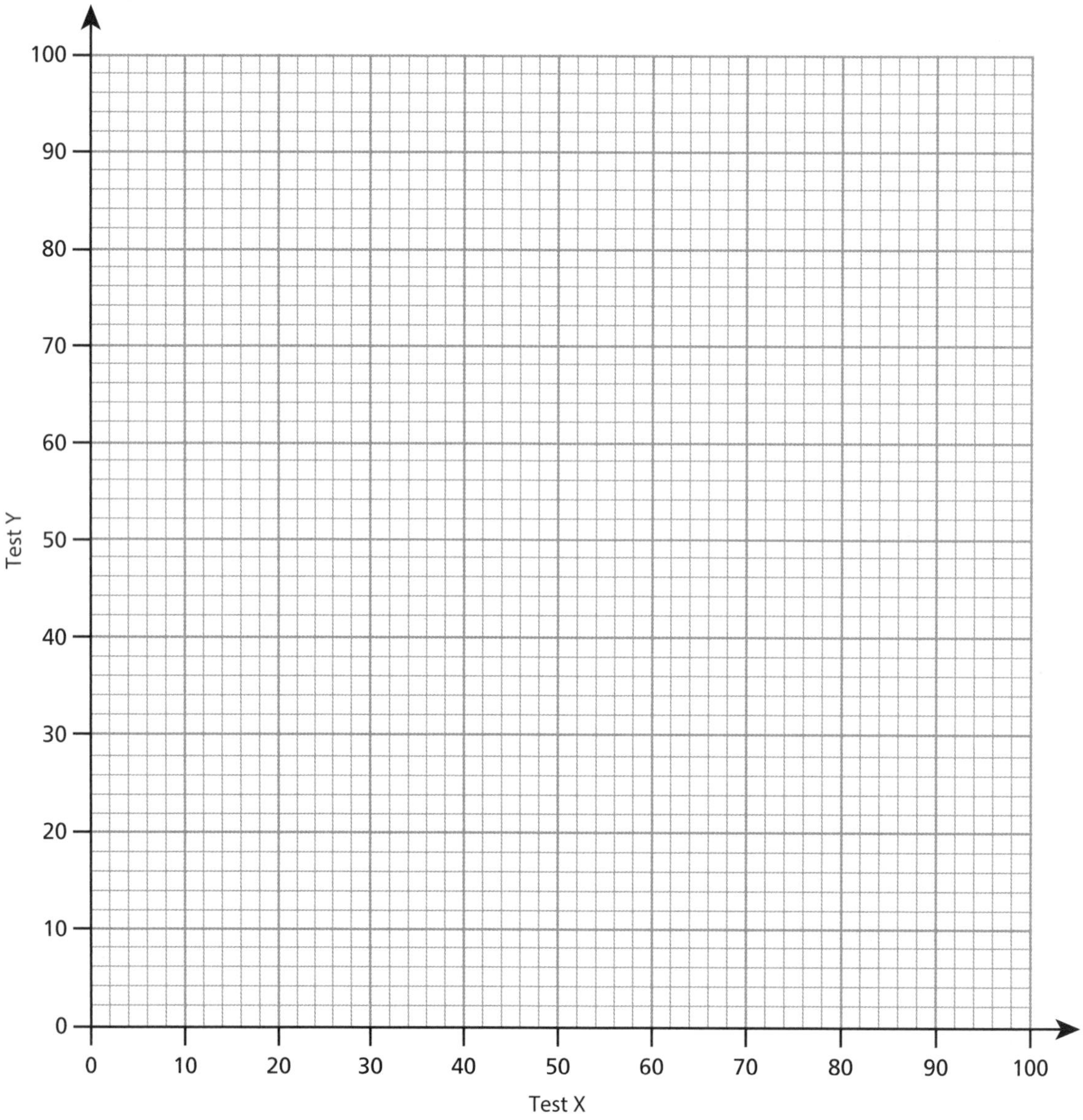

[graph: axes labelled Test Y (vertical, 0 to 100) and Test X (horizontal, 0 to 100)]

Test Y

Test X

[4]

b) Describe what the graph tells you.

....

.... [1]

(PS) **2** In a competition, there are four rounds before the final. To qualify for the final, competitors need a mean score of 50 or more points over the four rounds.

The table shows the scores of three competitors and their scores so far.

Competitor	Round 1	Round 2	Round 3	Round 4
A	54	38	50	
B	85	92		
C	41			

a) What is the lowest score that competitor A needs in Round 4 to qualify for the final?

.... [2]

b) Is it possible that competitor B could qualify for the final after three rounds? Show your working.

.... [1]

c) How many points does competitor C need to average over their last three rounds to qualify for the final?

.... [2]

(PS) **3** In a school, every student studies either Spanish or French. The pie chart shows the proportions for Year 8.

a) There are 210 students in Year 8.

How many study French?

.... [2]

Language studied by Year 8 students

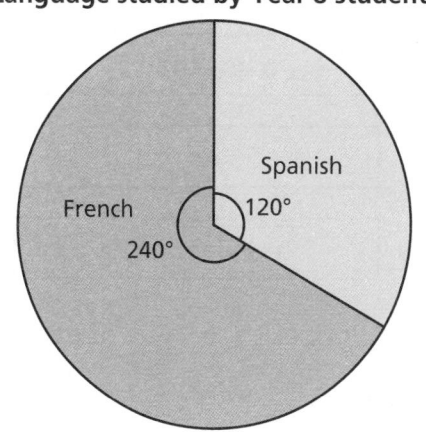

b) Exactly half of the students in Year 8 are boys. Ali says, "I can work out the number of boys who study French because it will be half of the total number who study French."

Is he correct? Give reasons for your answer.

....

.... [2]

Total Marks .... / 14

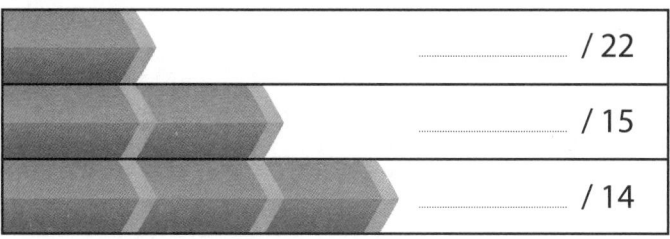

.... / 22

.... / 15

.... / 14

How do you feel about these skills?

(PS) (MR)

Green = Got it!
Orange = Nearly there
Red = Needs practice

Algebra

1 Expand:

  a) $3(x + 4)$ ....

  b) $5(x - 7)$ ....

  c) $7(2y + 4)$ ....

  d) $4(9y - 5)$ ....

  e) $a(a + 3)$ ....

  f) $b(b - 6)$ ....

  g) $2c(c + 2)$ ....

  h) $3d(2d - 5)$ .... **[8]**

2 Simplify:

  a) $9x + 4y - 6x$ ....

  b) $4r - 3s + 2s - r$ ....

  c) $5x - 3 + 7x - 2$ ....

  d) $3t + 5v - 4 + v - 7 + 2t$ .... **[4]**

MR **3** Draw lines to match each statement to the correct expression.

Statement	Expression
5 more than x	$5x$
5 less than x	$x + 5$
5 lots of x	$\dfrac{x}{5}$
x divided by 5	$x - 5$

[2]

4 Work out the value of each expression when $x = 4$, $y = -1$ and $z = 2$

  a) $x + y$ ....

  b) $x + y + z$ ....

  c) xz ....

  d) xy ....

  e) $\dfrac{x}{z}$ ....

  f) $\dfrac{x + y}{z}$ ....

  g) $z(x - y)$ ....

  h) $(2x + 3y)^2$ .... **[8]**

5 Simplify:

  a) $t \times t \times t \times t$ ....

  b) $w^2 \times w^5$ ....

  c) $x^4 \div x^2$ ....

  d) $\dfrac{a^2}{a}$ ....

  e) $\dfrac{b^7}{b^3}$ ....

  f) $\dfrac{5c^3}{c}$ ....

  g) $\dfrac{y^2 \times y^3}{y^4}$ ....

  h) $\dfrac{z^3 \times 5z^4}{z^2}$ .... **[8]**

6 The cost of printing 1 photo is 7 pence.

    a) Write an expression for the cost (in pence) of printing n photos.

                                                         [1]

    b) Write a formula for C, the cost (in pence) of printing n photos.

                                          $C =$ [1]

    c) Write a formula for P, the cost (in pounds) of printing n photos.

                                          $P =$ [1]

7 Dilip earns £18 per hour.

    Write a formula for E, the amount he earns for n hours' work.

                                          $E =$ [1]

8 $s = \dfrac{d}{t}$, where s is speed, d is distance and t is time.

    Mario runs 25 km in 3.5 hours.

    Use the formula to calculate Mario's speed in kilometres per hour.
    Give your answer to 1 decimal place. 🖩

                                        km/h   [2]

Total Marks   / 36

---

1 Expand:

    a) $-3(x + 6)$ **b)** $-2(x - 3)$

    c) $-(x + 8)$ **d)** $-2x(x - 4)$ [4]

2 Expand and simplify:

    a) $3(x + 2) + 4(x - 1)$ [2]

    b) $5(y - 7) - 2(y + 3)$ [2]

    c) $8(z - 2) - 3(z - 4)$ [2]

    d) $x(x + 3) - 2(x + 1)$ [2]

    e) $y(y - 2) + 3y(y + 4)$ [2]

    f) $3z(5z + 1) - z(4z + 7)$ [2]

3 Simplify fully:

    a) $\dfrac{10x^3}{2x}$ [1]

    b) $\dfrac{15x^7}{5x^3}$ [1]

    c) $\dfrac{3x \times 2x^4}{x^3}$ [1]

    d) $\dfrac{8x^2 \times 5x^5}{10x^4}$ [1]

Algebra

(MR) **4** Here is a rectangle.

Write an expression for:

a) the area of the rectangle

.... [1]

b) the total area of five of these rectangles

.... [1]

(MR) **5** The length of a rectangle is 3 cm greater than its width, w cm.

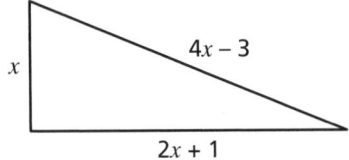

Write an expression for the area of the rectangle.

.... [2]

6 Factorise:

a) $4x + 8$ .... [1]

b) $5x - 20$ .... [1]

c) $10x + 15$ .... [1]

d) $21x - 14$ .... [1]

e) $x^2 - x$ .... [1]

f) $3x^2 + x$ .... [1]

g) $6x^2 - 4x$ .... [1]

h) $12x^2 + 15x$ .... [1]

(MR) **7** Write and simplify an expression for the perimeter of this triangle.

.... [2]

Total Marks .... / 34

1 Use the formula $y = (x - 2)^2 + k$ to work out the value of y:

a) when $x = 7$ and $k = 4$ $y = $ .... [1]

b) when $x = -2$ and $k = 11$ $y = $ .... [1]

(MR) **2** This shaded shape is made by cutting a rectangle out of a square.

Write an expression for the shaded area.

3x

x + 5

3x ☐ x

.... [3]

3 Simplify:

a) $(x^2)^2$ ....

b) $(y^3)^2$ ....

c) $(t^4)^2$ ....

d) $(2n)^2$ ....

e) $(5r^3)^2$ ....

f) $(3w^2)^3$ .... [6]

(PS) **4** $s = \dfrac{d}{t}$, where s is speed, d is distance and t is time. 🖩

a) A car travels at 65 km/h for 3 hours.

Work out the distance it travels.

.... km [2]

b) Betty cycles at 18 km/h.

How long does it take her to cycle 45 km?

.... hours [2]

5 Work out the value of each expression when $a = 5$, $b = 3$ and $c = -2$

a) a^2b ....

b) $(a - c)^2$ ....

c) $a + b + c^2$ ....

d) $\sqrt{20a}$ ....

e) $\sqrt{2(a + b)}$ ....

f) $ab + 4c$ ....

g) $(b - c)^2 + a$ ....

h) $\dfrac{ab}{c}$ .... [8]

(MR) **6** Write an expression for the area of this triangle.

2x

x² + 3

.... [2]

(FS) **7** Sal increases all the prices in her shop by 10%.

Write a formula to calculate the new price, n, for an item that costs £p.

.... [1]

Total Marks .... / 26

.... / 36

.... / 34

.... / 26

Fractions and Decimals

Calculators should **not** be used for any of the questions in this section.

1. Write each fraction as a decimal.

   a) $\frac{7}{10}$ ....   b) $\frac{1}{4}$ ....   c) $\frac{4}{5}$ ....   d) $\frac{9}{20}$ ....

   e) $\frac{11}{40}$ ....   f) $\frac{3}{8}$ ....   g) $\frac{38}{100}$ ....   h) $\frac{13}{200}$ ....   [8]

2. Write each decimal as a fraction in its simplest form.

   a) 0.3 ....   b) 0.09 ....

   c) 0.56 ....   d) 0.125 ....   [4]

3. Work out:

   a) $\frac{1}{5} \times \frac{1}{8}$ ....   b) $\frac{1}{5} + \frac{1}{8}$ ....

   c) $\frac{1}{10} - \frac{1}{12}$ ....   d) $\frac{1}{3} \div \frac{1}{4}$ ....

   e) $\frac{2}{7} \times \frac{5}{6}$ ....   f) $\frac{7}{9} - \frac{2}{5}$ ....

   g) $\frac{3}{5} \div \frac{3}{7}$ ....   h) $\frac{2}{9} + \frac{3}{4}$ ....   [8]

4. Write these in order, starting with the smallest:

   $\frac{3}{8}$ 20% $\frac{5}{12}$ 0.39 $\frac{17}{20}$ 0.876

   ....   [2]

5. Work out:

   a) 0.3×4 ....   b) 0.02×0.5 ....

   c) 0.24×0.2 ....   d) 3.5×0.4 ....   [4]

6. Work out:

   a) $1\frac{3}{4} + \frac{5}{8}$ ....   b) $2\frac{1}{4} - \frac{3}{4}$ ....

   c) $3\frac{1}{4} - 1\frac{5}{6}$ ....   d) $5\frac{2}{3} + 3\frac{1}{6}$ ....

   e) $4\frac{3}{10} + 1\frac{4}{5}$ ....   f) $3\frac{5}{7} + 2\frac{3}{5}$ ....   [6]

7. Work out:

   a) $1\frac{2}{5} \times 2\frac{1}{4}$ ....   b) $4\frac{5}{8} \div 1\frac{1}{2}$ ....

   c) $3\frac{3}{4} \div 1\frac{3}{5}$ ....   d) $6\frac{3}{5} \times 4\frac{1}{2}$ ....   [4]

Total Marks .... / 36

1 Write each fraction as a decimal to 3 decimal places.

a) $\frac{1}{3}$ .... b) $\frac{3}{7}$ .... c) $\frac{4}{9}$ .... d) $\frac{7}{9}$ .... [4]

(PS) 2 Work out the area of each shape in square inches.

a)

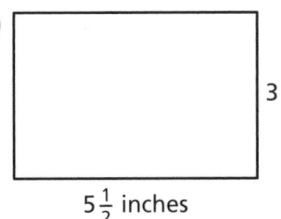

$3\frac{2}{3}$ inches

$5\frac{1}{2}$ inches

b)

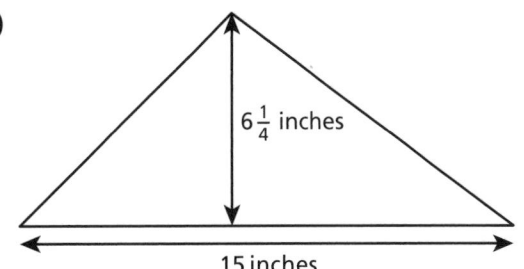

$6\frac{1}{4}$ inches

15 inches

.... inches² .... inches² [4]

3 Write the next three terms in each sequence as fractions or mixed numbers.

a) 100, 20, 4, ..., ..., ... [3]

b) 2, 3, $4\frac{1}{2}$, $6\frac{3}{4}$, ..., ..., ... [3]

4 The pie chart shows the favourite types of film among a group of students.

What fraction of the students chose these types of film? Write each fraction in its simplest form.

a) Romance

.... [1]

b) Thriller

.... [1]

c) Action

.... [1]

Favourite films

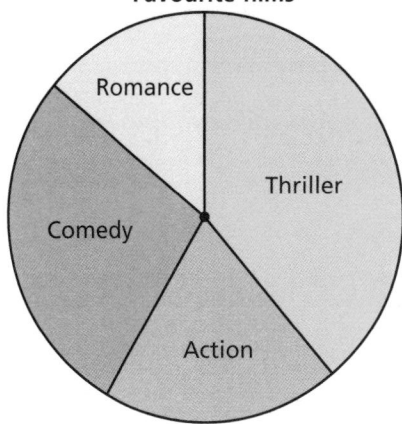

(MR) 5 $\frac{5}{12} \times 113\,832 = 47\,430$

a) Work out $\frac{1}{12}$ of $113\,832$ .... [1]

b) Work out $\frac{5}{6}$ of $113\,832$ .... [1]

(PS) 6 Write these in order, starting with the largest:

$1\frac{2}{5}$ 133% $1\frac{1}{3}$ 102%   1.2

.... [2]

Total Marks .... / 21

Fractions and Decimals

(PS) **1** The nth term of a sequence is $\dfrac{n+5}{2n}$

What is the value of the first term in the sequence that is less than 1?

.... [1]

2 Write the next three terms in each sequence.

a) $\dfrac{2}{5}, \dfrac{13}{20}, \dfrac{9}{10},$ .... , ... , ... [3]

b) $3\dfrac{5}{9}, 3\dfrac{1}{18}, 2\dfrac{5}{9},$ .... , ... , ... [3]

(PS) **3** Find the nth term of this sequence:

$5\dfrac{1}{4}, 5\dfrac{3}{4}, 6\dfrac{1}{4}, 6\dfrac{3}{4}, 7\dfrac{1}{4}, ...$

.... [2]

(PS) **4** Write these in ascending order:

$\dfrac{3}{8}$ -0.6 $-\dfrac{3}{4}$ $\dfrac{2}{3}$ -70%

.... [2]

(MR) **5** Find the value exactly halfway between $1\dfrac{3}{5}$ and $2\dfrac{5}{8}$

.... [2]

(MR) **6** Will the number $2\dfrac{1}{4}$ be in the sequence with nth term $\dfrac{n}{4} - 3$?

Explain how you know.

....

.... [2]

Total Marks .... / 15

.... / 36

.... / 21

.... / 15

Proportion

(FS) **1** 7 pens cost £8.05

Work out the cost of:

a) 1 pen   £ .... [1]

b) 3 pens   £ .... [1]

(PS) **2** 3 buses can take 96 passengers.

a) How many passengers can 5 buses take?   .... [1]

b) How many buses do you need for 500 passengers?   .... [1]

(PS) **3** Here are the ingredients to make 12 pancakes.

Makes 12 pancakes			
100 g flour	2 eggs	300 ml milk	1 teaspoon sunflower oil

How much of each ingredient do you need for the following?

a) 6 people   b) 30 people

.... flour   .... flour

.... eggs   .... eggs

.... milk   .... milk

.... teaspoons sunflower oil   .... teaspoons sunflower oil   [8]

(MR) **4** A car travels at 80 km/hour.

How many kilometres does it travel in:

a) 2 hours?   .... km [1]

b) $\frac{1}{2}$ hour?   .... km [1]

c) 15 minutes?   .... km [1]

(MR) **5** 3 people take 2 hours to build a wall.

How long will it take for:

a) 1 person to build an identical wall?   .... [1]

b) 6 people to build an identical wall?   .... [1]

c) 4 people to build an identical wall?   .... [1]

Total Marks .... / 18

Proportion

1 1 litre = 1.75 pints

Complete this conversion table for pints and litres.

Litres	0	1		8	
Pints			7		17.5

[5]

(MR) **2** In a factory, 1 machine fills 90 bottles in one hour.

a) How long would it take 3 of these machines
to fill 90 bottles between them? .... [1]

b) How long would it take 2 of these machines
to fill 270 bottles between them? .... [2]

(MR) **3** In a bakery, 1 machine fills 1500 pies in 2 hours.

Complete the table to show the times for different numbers of machines to fill 1500 pies.

Number of machines	1	2			10
Time to fill 1500 pies			30 minutes	15 minutes	

[5]

(PS) **4** Water drips steadily from a tap at a rate of 20 litres per hour.

a) How much water drips from the tap in 3 hours?

.... litres [1]

b) A 4-litre bucket is placed under the tap.

How long will it take for the bucket to
fill with water dripping from the tap? .... [1]

c) How much water drips from the tap in 1 week?

.... litres [1]

Total Marks .... / 16

1 8 km = 5 miles

Complete this conversion table for kilometres and miles.

Kilometres	0	1		16		32
Miles			5		15	

[6]

(MR) 2 Five cinema tickets cost £31.25

Write a formula for the cost C (in pounds) of n of these tickets.

_____ [2]

(MR) 3 The graph shows the cost of hiring a car.

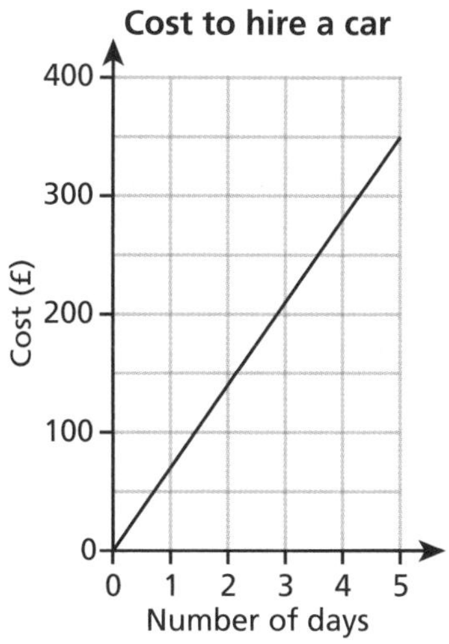

Cost to hire a car

a) How much does it cost to hire the car
for 8 days?

£ _____ [1]

b) Write a formula for the cost C (in pounds) of hiring this
car for d days.

_____ [1]

(PS) 4 1 kg = 2.2 pounds

Which is heavier, 8 kg or 18 pounds? Show your working.

_____

_____ [1]

(MR) 5 Draw lines to join each graph to the type of proportional relationship it shows between
x and y.

| Direct proportion | Inverse proportion | Not in proportion |

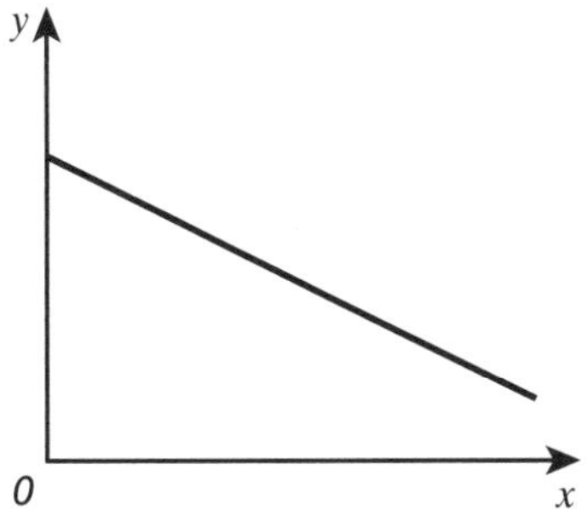

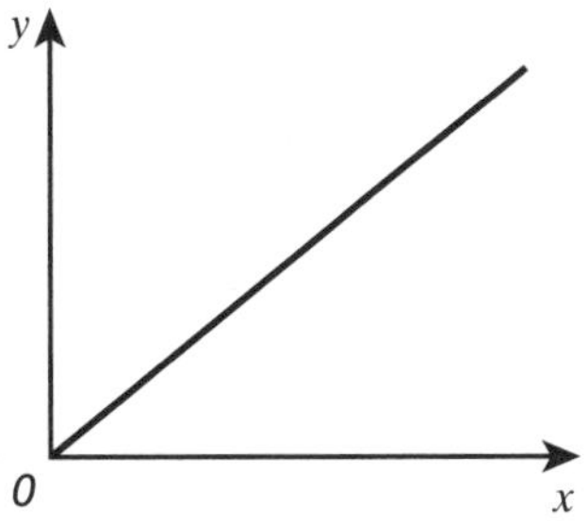

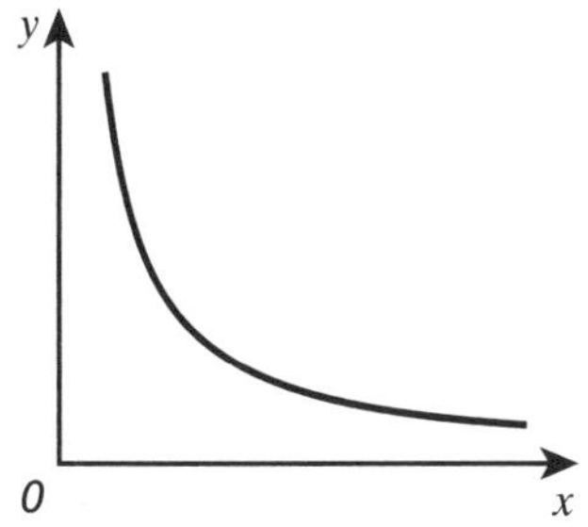

[2]

Total Marks _____ / 13

_____ / 18

_____ / 16

_____ / 13

Circles and Cylinders

1 $C = \pi d$ for the circumference of a circle. $A = \pi r^2$ for the area of a circle.

For each circle below, calculate:
- the circumference correct to 1 decimal place
- the area correct to 1 decimal place

a)

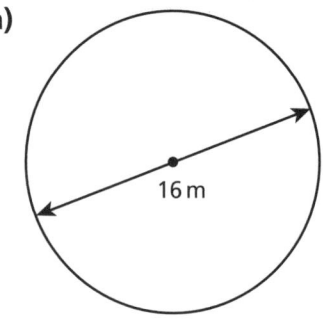

Circumference = _____ m

Area = _____ m² [2]

b)

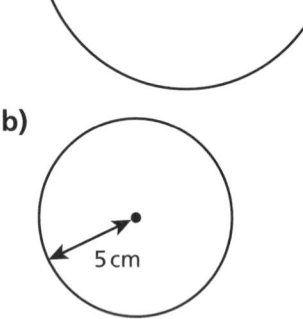

Circumference = _____ cm

Area = _____ cm² [2]

(PS) **2** Which shape has the longer perimeter and by how much: a circle of diameter 18 cm or a square of side 14 cm? Show your working.

_____

_____

_____

_____ [2]

(PS) **3** Ella jogs around this circular lake three times.

How many kilometres does she jog?
Give your answer to 1 decimal place.

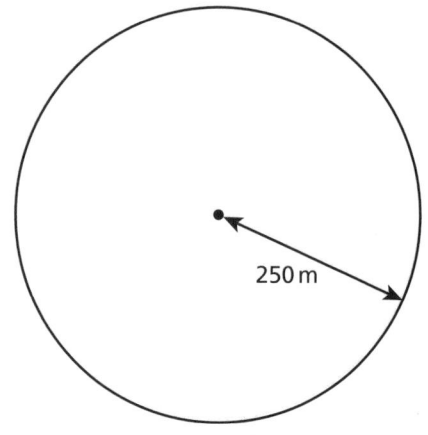

250 m

_____ km   [3]

4 a) A circle has an area of 314 cm².

Work out its radius. Use $\pi = 3.14$ 🚫

.... cm [2]

b) A circle has a circumference of 157 cm.

Work out its diameter. Use $\pi = 3.14$

.... cm [2]

(PS) **5** Work out the shaded area to 1 decimal place.

7 cm

4 cm

9 cm

.... cm² [3]

Total Marks .... / 16

(PS) **1** Calculate the area of this quarter circle to 1 decimal place.

6.8 cm

.... cm² [2]

(MR) **2** This garden is made from a rectangle and a semicircle. It is to be covered with gravel.
(PS) A £28.95 bag of gravel will cover 4 m².

(FS) What will be the total cost of buying enough bags to cover the entire garden with gravel?

7 m

12 m

£ .... [5]

Circles and Cylinders

3 Work out the volume of this cylinder. Give your answer as a multiple of π.

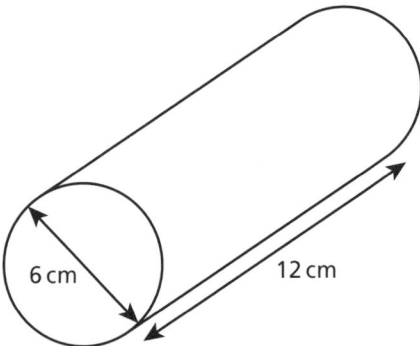

6 cm   12 cm

_____ cm³ [3]

(MR) (PS) **4 a)** Work out the area of a circle with diameter 10 cm.
Give your answer as a multiple of π.

_____ cm² [2]

b) This circle has diameter 10 cm.

Work out the shaded area, giving your answer in terms of π.

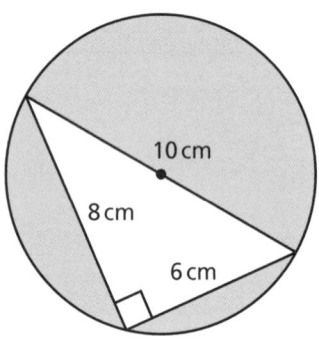

10 cm

8 cm

6 cm

_____ cm² [2]

(PS) (FS) **5** These two pieces of fabric are a semicircle and a quarter circle. They have ribbon sewn completely around their edges. The ribbon costs 5p per centimetre.

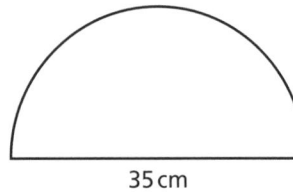   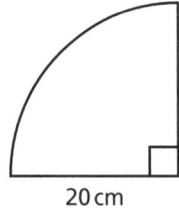

35 cm   20 cm

Work out the total cost of the ribbon used.

£ _____ [6]

(PS) **6** A cylindrical container is shown below.

a) Work out the volume of the container, giving your answer in cm³ to 3 decimal places.

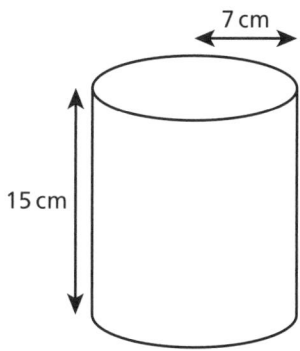
7 cm

15 cm

.... cm³ [2]

b) 1 ml = 1 cm³

Work out the capacity of the cylinder in litres.
Give your answer to 2 significant figures.

.... litres [1]

Total Marks .... / 23

(MR) **1** In the diagram, the radius of the smaller circle is 5 cm and the radius of the larger circle is 10 cm.

The circumference of the smaller circle is 31.4 cm.

Without using the formula for the circumference of a circle, work out the circumference of the larger circle.

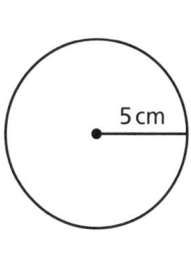
5 cm

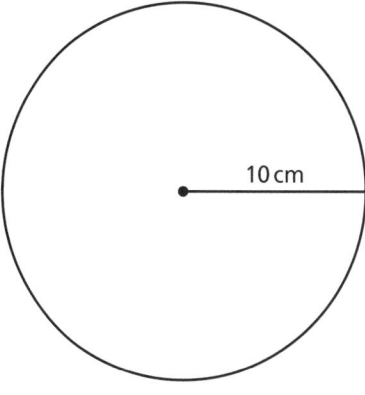
10 cm

.... cm [2]

(MR) (PS) **2** A swimming pool has diameter 18 m. There is a 1-metre wide path around the swimming pool.

Work out the area of the path, giving your answer in terms of π.

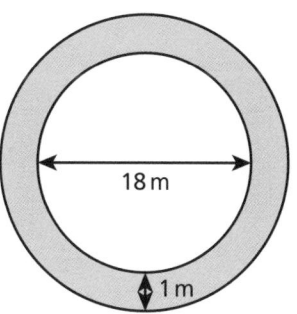
18 m

1 m

.... m² [4]

Circles and Cylinders

(MR) **3** The volume of this cylinder is $80\pi\,\text{cm}^3$. Another cylinder has twice the radius of this cylinder and the same height.

What is the volume of the other cylinder? Give your answer in terms of π.

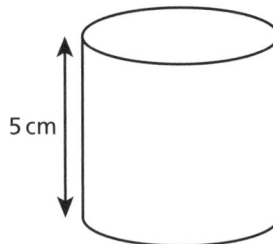

5 cm

.... cm^3 [5]

(MR) (PS) **4** A cricket pitch is 20.1 m long. A cricket ball has diameter 7.2 cm. The ball is rolled from one end of the pitch to the other end.

How many **complete** rotations will the ball make?

.... [3]

(MR) (PS) **5** This is a running track. It has semicircular ends.

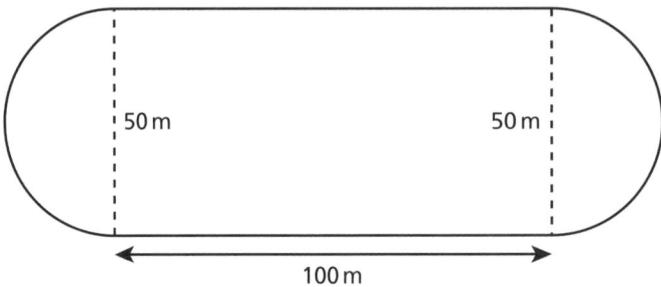

50 m   50 m

100 m

a) Work out the perimeter of the running track to 1 decimal place.

.... m [2]

b) An athlete runs at a speed of 7.5 metres per second.

How long will the athlete take to run two laps?
Give your answer to 1 decimal place.

.... s [2]

6 Work out the shaded area to 1 decimal place.

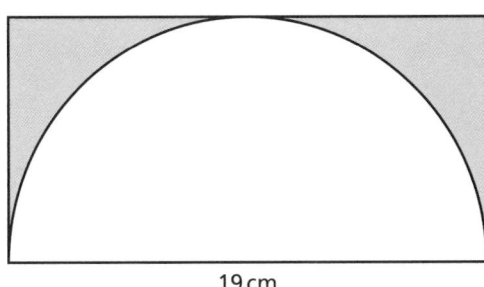

19 cm

_____ cm² [4]

7 Sofia has a cylindrical paddling pool as shown.

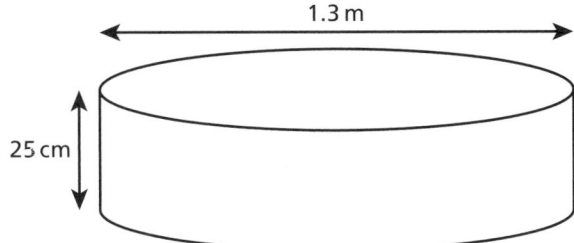

1.3 m

25 cm

The paddling pool contains 0.265 m³ of water.

Work out the height of the water in centimetres to the nearest whole number.

_____ cm [3]

Total Marks _____ / 25

_____ / 16

_____ / 23

_____ / 25

Equations and Formulae

1 Solve the equations.

a) $4 + a = 7$ **b)** $b + 7 = 11$ **c)** $c + 5 = 2$

$a =$ _____ [1] $b =$ _____ [1] $c =$ _____ [1]

d) $d - 4 = 0$ **e)** $e - 9 = 10$ **f)** $f - 7 = -1$

$d =$ _____ [1] $e =$ _____ [1] $f =$ _____ [1]

2 Solve the equations.

a) $5a = 30$ **b)** $4b = 28$ **c)** $8c = 80$

$a =$ _____ [1] $b =$ _____ [1] $c =$ _____ [1]

d) $12d = 6$ **e)** $6e = -36$ **f)** $4f = -3$

$d =$ _____ [1] $e =$ _____ [1] $f =$ _____ [1]

g) $\frac{g}{2} = 8$ **h)** $5h = 0$ **i)** $\frac{x}{4} = -4$

$g =$ _____ [1] $h =$ _____ [1] $x =$ _____ [1]

3 The angles in a triangle are a, $2a$ and $3a$.

Work out the size of the largest angle.

_____ ° [2]

4 Solve the equations.

a) $5a + 1 = 21$

b) $3b - 4 = 23$

c) $2c + 1 = 10$

$a =$ .... [2] $b =$ .... [2] $c =$ .... [2]

d) $4d - 7 = 13$

e) $2e + 6 = 6$

f) $5f + 2 = -3$

$d =$ .... [2] $e =$ .... [2] $f =$ .... [2]

g) $3g + 24 = 15$

h) $\frac{h}{2} + 1 = 15$

i) $\frac{x}{3} - 2 = 4$

$g =$ .... [2] $h =$ .... [2] $x =$ .... [2]

(PS) **5** The length of this rectangle is double the width.
The width is x cm. The perimeter is 36 cm.

a) Write down an expression for the length of the rectangle.

x cm

.... cm [1]

b) Set up and solve an equation to work out the width of the rectangle.

.... cm [2]

6 Make t the subject of each formula.

a) $s = t + 10$ .... [1]

b) $v = t - 4.5$ .... [1]

c) $x = 5t$ .... [1]

d) $y = \frac{1}{2}t$ .... [1]

7 Make x the subject of each formula.

a) $y = 2x + 3$ .... [2]

b) $T = x + y + z$ .... [2]

c) $5x - 4 = p$ .... [2]

d) $w = 8 - 6x$ .... [2]

Total Marks .... / 50

Equations and Formulae

1 Solve the equations.

a) $4(x + 5) = 24$

$x =$ .... [3]

b) $4(x + 5) = 3(x + 8)$

$x =$ .... [3]

c) $4(x + 5) = 2(x - 7)$

$x =$ .... [3]

2 Solve the equations.

a) $5x = x + 12$

$x =$ .... [2]

b) $5x + 7 = x + 19$

$x =$ .... [2]

c) $6x - 2 = 2x + 10$

$x =$ .... [2]

3 Solve the equations.

a) $3(a + 2) = 15$

$a =$ .... [3]

b) $3(b + 2) = b$

$b =$ .... [3]

c) $3(c + 2) = c + 15$

$c =$ .... [3]

d) $\frac{1}{2}(d - 3) = 20$

$d =$ .... [3]

e) $\frac{1}{2}(e - 3) = e$

$e =$ .... [3]

f) $\frac{1}{2}(f - 3) = f + 20$

$f =$ .... [3]

(PS) **4** The diagram shows an equilateral triangle and a regular pentagon.
The perimeters of the shapes are equal.

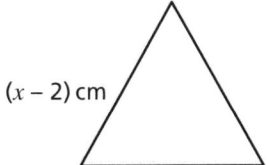 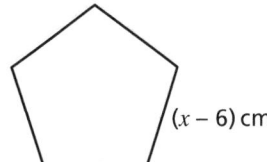

$(x - 2)$ cm $(x - 6)$ cm

a) Set up and solve an equation to work out the value of x.

$x =$ .... [4]

b) Work out the perimeter of the triangle.

.... cm [2]

5 Rearrange each formula to make x the subject.

a) $y = x + 4$

.... [1]

b) $2y = x + 4$

.... [1]

c) $y = 2x + 4$

.... [2]

d) $y = 2x - 4$

.... [2]

e) $y = 2x + 3w$

.... [2]

(MR) **6** The two parallel sides of a trapezium are 7 cm and 3 cm.

(PS) Work out the height of the trapezium if its area is 40 cm².

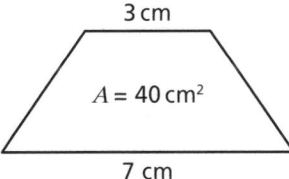

3 cm

$A = 40$ cm²

7 cm

.... cm [2]

Total Marks .... / 49

Equations and Formulae

1 Solve the equations.

a) $\frac{x+4}{2} = 6$

b) $\frac{x-4}{2} = 11$

c) $\frac{x-4}{5} = 7$

$x =$ .... [2] $x =$ .... [2] $x =$ .... [2]

(MR) **2** Here are two straight lines. The lines are the same length.

The first line is divided into three equal parts. The second line is divided into four equal parts.

$(x + 5)$ cm

$2x$ cm

Show that the length of each line is 24 cm.

....

....

....

.... [5]

3 Solve the equations.

a) $\frac{1}{3}(x + 4) = 8$

b) $\frac{2}{3}(x + 4) = 8$

c) $\frac{2}{3}(x + 4) = x + 8$

$x =$ .... [2] $x =$ .... [3] $x =$ .... [3]

(PS) (MR) **4** The surface area of the cuboid is 128 cm².

Find the value of x. Show your working.

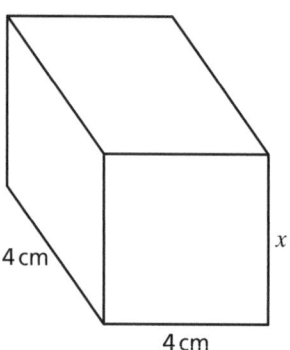

4 cm

x

4 cm

$x =$ .... cm [3]

(PS) **5** A large bottle of juice will fill 6 glasses with 10 cl left over.
A small bottle of juice will fill 2 glasses with nothing left over.
The small bottle holds 1.1 litres less than the large bottle.

Work out how much each bottle holds.

Small bottle holds cl

Large bottle holds cl   [5]

(MR) **6** The area of the rectangle and the area of the square are equal.

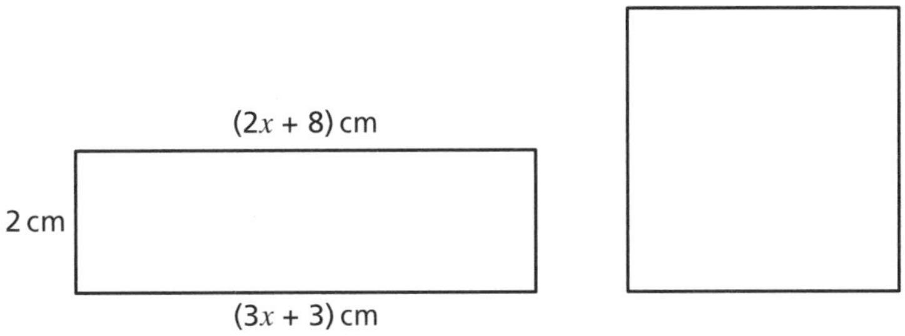

$(2x + 8)$ cm

2 cm

$(3x + 3)$ cm

Work out the length of one side of the square.

.... cm   [6]

Total Marks .... / 33

.... / 50

.... / 49

.... / 33

How do you feel about these skills?
(PS) (MR) Green = Got it! Orange = Nearly there Red = Needs practice

Statistics: Average and Range

1. For each set of data, calculate the range.

   a) 15, 1, 2, 8, 7

   .... [1]

   b) –5, –2, 8, 3, 6

   .... [1]

2. Here are the scores of 11 students in a test.

9	8	10	1	8	4	10	6	6	9	6

   a) Write down the mode.

   .... [1]

   b) Work out the median.

   .... [2]

   c) Work out the mean.

   .... [2]

   d) Work out the range.

   .... [1]

3. The temperature at midday (in °C) of 12 towns is shown.

6	12	8	8	9	10	5	7	5	6	6	6

   a) Which temperature is the mode?

   .... °C [1]

   b) What is the range of the temperatures?

   .... °C [1]

   Total Marks .... / 10

(MR) **1** a) Work out the median of 8, 11, 6, 4, 7, 0, 5, 4

.... [2]

b) Use your answer to part a) to write down the median of
108, 111, 106, 104, 107, 100, 105, 104

.... [1]

(MR) **2** The table shows the times 30 students take to complete a task.

Time, t (minutes)	Frequency
$0 < t \leqslant 10$	16
$10 < t \leqslant 20$	8
$20 < t \leqslant 30$	6

For each statement, decide if it is:

Definitely true   **Definitely false**   **Could be true or false**

a) 8 students took less than 5 minutes to complete the task.

.... [1]

b) A student who took 10 minutes to complete the task is in the class $10 < t \leqslant 20$

.... [1]

c) The range of the times taken is 20 minutes.

.... [1]

d) The slowest time taken is more than 20 minutes.

.... [1]

(MR) **3** The data shows the mean score and the range of scores for two quiz teams.

	Team A	Team B
Mean	18	16
Range	6	8

Which is the better team? Give **two** reasons for your answer.

....

....

.... [2]

Statistics: Averages and Range

(MR) **4** Akeel records the number of goals scored by his team in each of the last 20 matches.

Number of goals	Number of games
0	5
1	7
2	3
3	4
4	0
5	1

a) Work out the mean number of goals scored.

.... [3]

b) In the next match, Akeel's team scores two goals.

Is the mean for all 21 matches more or less than the mean for the 20 matches?
Explain your answer.

....

.... [1]

Total Marks .... / 13

(MR) **1** a) Work out the range of 34, 52, 85, 26, 18, 49

.... [2]

b) Use your answer to part a) to write down the range of 1034, 1052, 1085, 1026, 1018, 1049

.... [1]

(PS) **2** The median of these five numbers is 5

| $x + 1$ | | $x + 4$ | | $x + 8$ | | $x + 1$ | | $x + 7$ |

a) Work out the value of x.

$x =$ .... [2]

b) Work out the value of the mode. .... [1]

3 The profits made each day for July and August by a shop are summarised in the table.

Profit, P (£)	$0 < P \leqslant 50$	$50 < P \leqslant 100$	$100 < P \leqslant 150$	$150 < P \leqslant 200$
Frequency (July)	8	17	4	2
Frequency (August)	15	12	1	3

a) Complete the dual bar chart to show the profits for July and August.

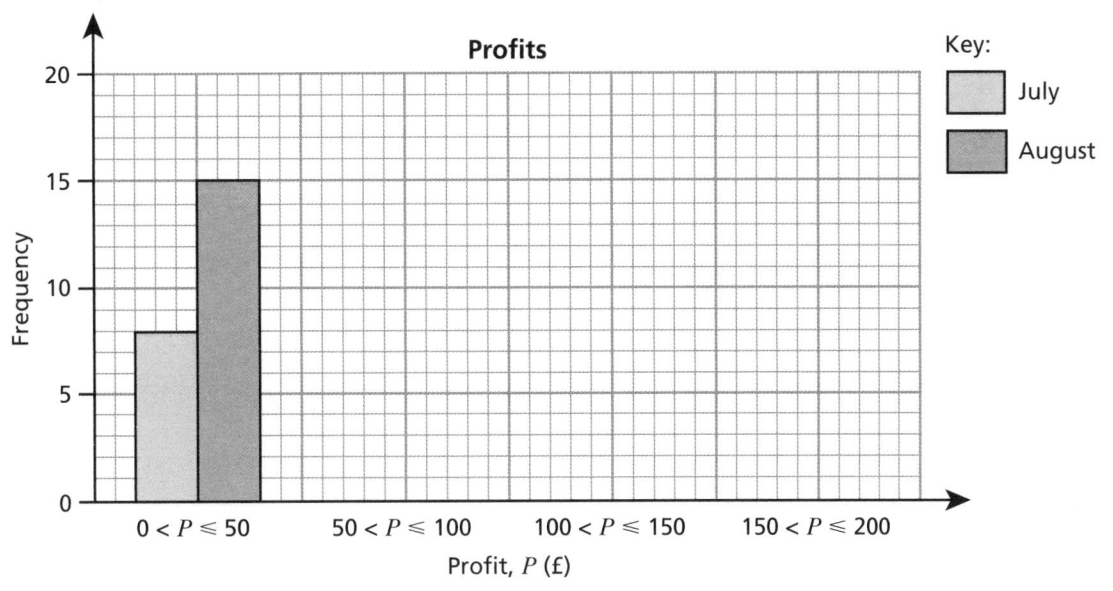

[3]

b) Compare the profits for July and August.

Which month do you think was more profitable?

....

....

.... [2]

c) Is it possible to compare the ranges of the profits for July and August? Give a reason for your answer.

....

.... [1]

Statistics: Averages and Range

(PS) **4** A competition has five rounds. The scores in each round are shown for the three players. The highest mean score wins. The player names have been left off the scorecard.

Round	1	2	3	4	5
Player A	6	8	8	7	6
Player B	4	10	3	9	2
Player C	6	7	7	7	6

a) Jo says, "I was the most consistent."

Imran says, "I won the competition."

Bob says, "I was the other player."

Can you identify the players from their statements? Show your working.

Jo is player ....   Imran is player ....   Bob is player ....   [4]

b) In what position did Bob finish?

.... [1]

(PS) **5** The tables show the times that two workers, A and B, start and finish each job on one day.

Worker A

Job number	1	2	3	4	5	6
Start time	9.00 am	9.35 am	10.10 am	11.55 am	1.00 pm	3.05 pm
Finish time	9.30 am	10.05 am	11.50 am	12.15 pm	3.00 pm	4.00 pm
Time taken, t (minutes)						

Worker B

Job number	1	2	3	4	5	6	7
Start time	8.00 am	8.25 am	9.10 am	10.55 am	1.00 pm	1.35 pm	2.10 pm
Finish time	8.20 am	9.05 am	10.50 am	11.55 am	1.30 pm	2.05 pm	3.30 pm
Time taken, t (minutes)							

a) Complete the tables for times taken for each job. [4]

b) Work out the range of times taken for each worker.

Worker A: _____ minutes

Worker B: _____ minutes [2]

c) Which worker spent the greater total time on jobs? Show your working.

_____ [2]

d) Complete the grouped frequency tables for workers A and B.

Worker A

Time taken, t (minutes)	Frequency
$0 < t \leqslant 30$	3
$30 < t \leqslant 60$	
$60 < t \leqslant 90$	
$90 < t \leqslant 120$	

Worker B

Time taken, t (minutes)	Frequency
$0 < t \leqslant 30$	
$30 < t \leqslant 60$	
$60 < t \leqslant 90$	
$90 < t \leqslant 120$	

[4]

e) Complete the frequency diagram for the times taken by worker A.

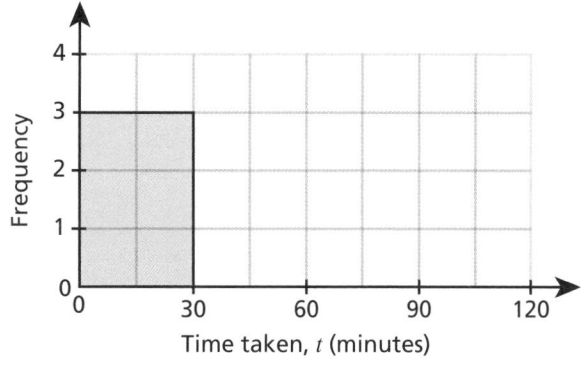

[2]

Total Marks _____ / 31

Answers

Pages 4–7: Working with Numbers

1. a) $1^3 = 1 \times 1 \times 1 = 1$ **[1]**   b) $8^2 = 8 \times 8 = 64$ **[1]**   c) 6 **[1]**
   d) 3 **[1]**   e) 4 **[1]**   f) 125 **[1]**   g) 16 **[1]**   h) 2 **[1]**

2.
Use the order of operations:			
Brackets	Indices (powers and roots)	Division or Multiplication	Addition or Subtraction

   a) $2 \times 8 - 3 + 5 = 16 - 3 + 5 = 13 + 5 = 18$ **[1]**
   b) $\frac{7+3}{5} - 2 = \frac{10}{5} - 2 = 2 - 2 = 0$ **[1]**

   The dividing line 'acts as a bracket', so you work out 7 + 3 first.

   c) $4 + 3 \times 5 = 4 + 15 = 19$ **[1]**
   d) $6^2 + 2 \times 5 = 36 + 10 = 46$ **[1]**
   e) $8 + 5(4 + 7) - 6 = 8 + 5 \times 11 - 6 = 8 + 55 - 6 = 57$ **[1]**
   f) $2^3 (15 - 6) + 2 \times 3 = 8 \times 9 + 2 \times 3 = 72 + 6 = 78$ **[1]**
   g) $\sqrt{49} + 2(3 + 5) - 4 = 7 + 2 \times 8 - 4 = 7 + 16 - 4 = 19$ **[1]**
   h) $\frac{40+\sqrt{25}}{3^2} = \frac{40+5}{9} = \frac{45}{9} = 5$ **[1]**

3. Simplify a ratio by dividing both numbers by the same number (a common factor).

   a) $\div 2 \begin{pmatrix} 2 : 8 \\ 1 : 4 \end{pmatrix} \div 2$ **[1]**   b) 2 : 1 **[1]**   c) 4 : 5 **[1]**
   d) 5 : 2 **[1]**   e) 2 : 3 **[1]**   f) 2 : 5 **[1]**

4. 3 boxes at £10.50 each = £31.50
   50 × 3 = 150 badges
   Selling 150 badges at 40p: 150 × 0.4 = £60 **[1]**
   Profit £60 − £31.50 = £28.50 **[1]**

5. 5 apples: 5 × 24 = 120p = £1.20
   4 bananas: 4 × 17 = 68p
   Total: £1.88 **[1]**
   Cost of strawberries = £5 − £1.02 − £1.88 = £2.10 **[1]**

6. a) −9 **[1]**   b) −32 **[1]**   c) 35 **[1]**   d) −9 **[1]**   e) −6 **[1]**   f) 6 **[1]**
   g) $(-5)^2 = -5 \times -5 = 25$ **[1]**   h) $(-2)^3 = -2 \times -2 \times -2 = 4 \times -2 = -8$ **[1]**

7. a) −5 **[1]**   b) 4 **[1]**   c) −9 **[1]**   d) 5 + 3 = 8 **[1]**
   e) $-6 - 2 = -8$ **[1]**   f) $-2 + 5 = 3$ **[1]**   g) 0 **[1]**   h) 26 **[1]**

1. 23, 29, 31, 37 **[2]**
   [1 mark for three correct and no incorrect values]

   A prime number has exactly two factors.

2. a)   b)
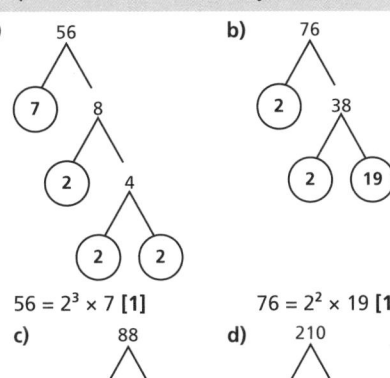

   $56 = 2^3 \times 7$ **[1]** $76 = 2^2 \times 19$ **[1]**

   c)   d)
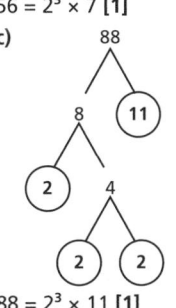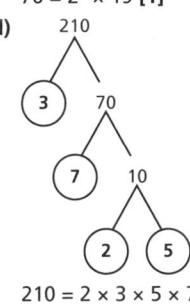

   $88 = 2^3 \times 11$ **[1]** $210 = 2 \times 3 \times 5 \times 7$ **[1]**

3. a) HCF = 2 **[1]**   LCM = 140 **[1]**
   b) HCF = 14 **[1]**   LCM = 84 **[1]**

   To find the HCF, you can list all the factors of both numbers and find the highest one they have in common. To find the LCM, you can list multiples of both numbers until you find the lowest one they have in common.

4. a) 1 hour 54 minutes **[1]**
   b) 2102 **[1]**
5. HCF = $2^2 \times 5 = 20$ **[1]**
   LCM = $2^2 \times 3 \times 5 \times 7 = 420$ **[1]**

   The HCF is the product of all the numbers in the intersection of the Venn diagram. The LCM is the product of all the numbers in the Venn diagram.

6. White : Red
      4 : 1
     20 : 5 **[1]**
   Total 20 + 5 = 25 litres of pink paint **[1]**

   Jack cannot use all the 7 litres of red paint, because then he would need 28 litres of white paint, and he only has 20 litres.

7. LCM of 25 seconds and 40 seconds = 200 seconds **[1]**
   200 seconds = 3 minutes 20 seconds **[1]**

1. To find the distance from Copenhagen to Madrid, read across from Copenhagen and down from Madrid.

Rome				
1353	**Madrid**			
1068	2846	**Kyiv**		
1530	2061	1320	**Copenhagen**	
1088	2380	1481	2142	**Athens**

   a) 2061 km **[1]**
   b) 1068 km **[1]**
   c) 2380 + 1353 = 3733 **[2]**
   [1 mark for either 2380 or 1353 if final answer is incorrect]

2. a) For the HCF, find the product of the factors they have in common: 2 × 3 = 6 **[1]**
   For the LCM, multiply the highest powers of each factor:
   $2^2 \times 3^2 \times 5 \times 7 = 1260$ **[1]**
   b) $150 = 2 \times 3 \times 5^2$ $220 = 2^2 \times 5 \times 11$
   HCF = $2 \times 5 = 10$ **[1]**
   LCM = $2^2 \times 3 \times 5^2 \times 11 = 3300$ **[1]**

3. a) −1 **[1]**   b) −10 **[1]**   c) 6 × 5 = 30 **[1]**
   d) $\frac{-4}{2} = -2$ **[1]**   e) $\frac{100}{25} = 4$ **[1]**   f) $\frac{12-2-20}{-5} = \frac{-10}{-5} = 2$ **[1]**

4. 12 jars cost £4 **[1]**
   3 kg fruit 3 × £3.40 = £10.20 **[1]**
   3 kg sugar 3 × £0.78 = £2.34
   Total cost £4 + £10.20 + £2.34 = £16.54 **[1]**
   Total sales 12 × £2.30 = £27.60 **[1]**
   Profit = £27.60 − £16.54 = £11.06 **[1]**

   There are different ways of solving this. This is just one method.

5.   400 g

A	A	A	D	D	D	D	D

   50 g
   400 g ÷ 8 = 50 g **[1]**
   Alix gets 3 × 50 g = 150 g
   Dan gets 5 × 50 g = 250 g **[1]**

6.

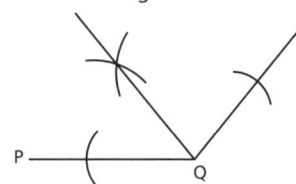

£16 ÷ 4 [1]
Jo gets £4 [1]

Pages 8–14: Geometry

1.

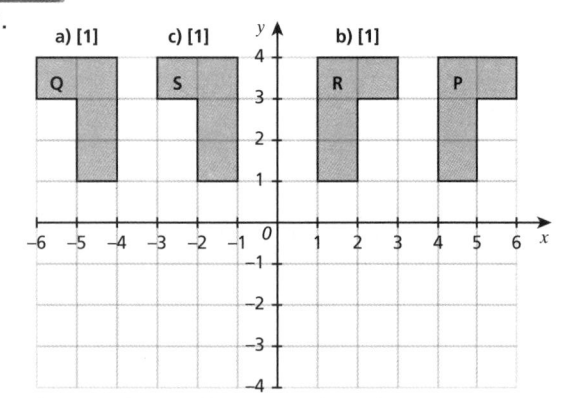

d) A translation of 3 units to the right [1]

2. Yes, correct and 180° − 75° − 30° or 75° [1]
So two equal angles or 75°, 75°, 30°, so isosceles [1]

3. a) h [1]   b) f [1]   c) g [1]

4. rectangle; square [2] [1 mark for one correct shape and no
incorrect shapes or two correct shapes and one incorrect shape]

5. $a = 69°$ [1] $b = 111°$ [1] $c = 67°$ [1] $d = 113°$ [1]
$e = 67°$ [1] $f = 43°$ [1] $g = 43°$ [1] $h = 94°$ [1]

6. a) (4, 1) (5, 1) (4, 4) [2]
   [1 mark for one correct pair of coordinates]
   b) (4, −1) (5, −1) (4, −4) [2]
   [1 mark for one correct pair of coordinates]
   c) (−2, 1) (−1, 1) (−2, −2) [2]
   [1 mark for one correct pair of coordinates]
   d) 180° [1] centre (−2, 1) [1]

Use tracing paper if needed.

1. Ratio of angles A : B : C : D is 4 : 3 : 2 : 1 [1]
Angles in a quadrilateral = 360° [1]
360 ÷ 10 = 36° [1]

2. a) The arcs should be equal sized, centred on A and B. [1]
Line should be cut into two equal parts. [1]

   b) The diagonals are perpendicular to each other. [1]

3. Quadrilateral made from two triangles [1]
So sum of angles = 2 × 180° = 360° [1]

4.

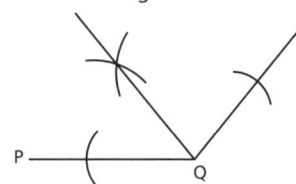

[1 mark for arcs; 1 mark for
angle bisector line]

5. $a = 70°$ [1] $b = 35°$ [1] $c = 75°$ [1] $d = 83°$ [1] $e = 30°$ [1]
6. $a = 68°$ [1]
Reason: Angles on a straight line add up to 180°. [1]
$b = 68°$ [1]
Reason: Alternate (and therefore equal) to angle a. [1]
$c = 55°$ [1]
Reason: Angles on a straight line add up to 180°. [1]
$d = 125°$ [1]
Reason: Corresponding to the marked 125° angle. [1]
$e = 38°$ [1]
Reason: Angles in a quadrilateral add up to 360°. [1]
$f = 38°$ [1]
Reason: Vertically opposite to angle e. [1]

$g = 71°$ [1]
Reason: Angles in a triangle add up to 180° and base angles
in an isosceles triangle are equal. [1]
7. Angle STQ = 65° (angles on a straight line add up to 180°) [1]
Angle TQR (64°) should equal STQ (65°) if alternate angles
but do not so PR and SU are not parallel [1]

There are other ways of explaining, for example using
corresponding angles.

8. Angle BCD = 110° (angles on a straight line add up to 180°) [1]
Angle BDC = 35° (angles in isosceles triangle) [1]
Angle ABD = 35° (alternate angles are equal) [1]
Angle ADB = 35° (angles in triangle ABD add up to 180°) [1]
Triangle ABD is isosceles (angle ABD = angle ADB) [1]

1. 90° [1] clockwise [1] about the origin or (0, 0) [1]
[Accept 270° anti-clockwise about the origin or (0, 0)]
2.

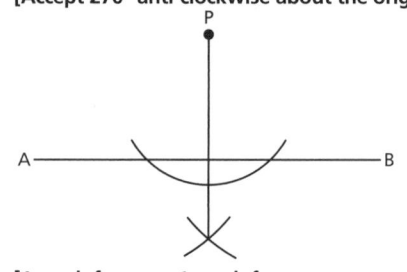

[1 mark for arcs; 1 mark for correct perpendicular line]
3. Angle CBD = 60° (corresponding) [1]
Angle FBE = 40° (alternate) [1]
Angle EBD = 180 − 40 − 60 − 40 [1]
= 40° (angles on straight line) [1]
Two angles equal so isosceles. [1]
4.

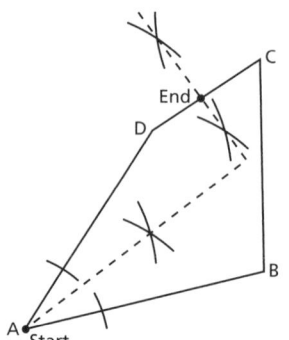

[5]
[1 mark for the angle bisector; 1 mark for its construction lines;
1 mark for the perpendicular bisector; 1 mark for the
construction lines; 1 mark for the complete route]
5. $x = a$ Reason: Alternate angles are equal.
$y = c$ Reason: **Alternate angles are equal.** [2]
$x + b + y = 180°$ Reason: **Angles on a straight line add up
to 180°** [2]
So $a + b + c = 180°$ [1]
So angles in a triangle add up to **180°** [1]

6. AB and PQ are the diagonals of a rhombus [1]. The diagonals of a
rhombus cross at right angles [1] and cut each other in half [1].
7. Angle QSR = a (isosceles triangle) [1]
$5a = 180°$ so $a = 36°$ [1]
Angle PSQ = 45° (isosceles triangle) [1]
Angle PSR = 36° + 45° = 81° [1]

Pages 15–17: Percentages

1. a) 4.5 cm [1]   b) 250 ml [1]   c) £3 [1]   d) 1.5 kg [1]
   e) 40 m [1]   f) 5.4 litres [1]
2. a) 12% [1]   b) 85% [1]   c) 67.5% [1]   d) 40% [1]
   e) 12.5% [1]   f) 62.5% [1]

To write a fraction as a percentage, convert the fraction to a
decimal, then to a percentage. e.g. $\frac{28}{80}$ = 28 ÷ 80 = 0.35 = 35%

3. a) 0.15 [1] b) 0.26 [1] c) 1.2 [1] d) 0.02 [1] e) 0.083 [1]
   f) 1.07 [1]
4. Boxes joined as follows:

Work out 30% of £36	to	0.3 × 36
Work out 24% of £36	to	0.24 × 36
Work out 95% of £36	to	0.95 × 36
Work out 3% of £36	to	0.03 × 36
Work out 2.4% of £36	to	0.024 × 36
Work out 110% of £36	to	1.1 × 36

   [4]
   [3 marks for three correct; 2 marks for two correct; 1 mark for one correct]

> To calculate a percentage of a quantity, convert the percentage to a decimal and multiply. e.g. 20% of 60 kg = 0.2 × 60 = 12 kg

5. a) 1.1 × £56 = £61.60 [1]

> Increase by 10% means find 110%

   b) 1.3 × 28 g = 36.4 g [1]
   c) 1.5 × 450 ml = 675 ml [1]
   d) 1.05 × 46 m = 48.3 m [1]
6. a) 0.9 × £25 = £22.50 [1]

> Decrease by 10% means find 100% − 10% = 90%

   b) 0.75 × 50 m = 37.5 m [1]
   c) 0.6 × 32 kg = 19.2 kg [1]
   d) 0.95 × 20 litres = 19 litres [1]

1. a) 66.7% [1] b) 71.4% [1] c) 44.4% [1] d) 73.5% [1]
   e) 81.7% [1] f) 78.6% [1]

> To write a fraction as a percentage, convert the fraction to a decimal, then to a percentage. Example: $\frac{3}{11}$ = 3 ÷ 11 = 0.27272727… = 27.272727…% = 27.3% (1 d.p.)

2. Boxes joined as follows:

Increase £120 by 10%	to	1.1 × 120
Decrease £120 by 10%	to	0.9 × 120
Increase £120 by 4%	to	1.04 × 120
Decrease £120 by 4%	to	0.96 × 120
Increase £120 by 40%	to	1.4 × 120
Decrease £120 by 40%	to	0.6 × 120

   [4]
   [3 marks for three correct; 2 marks for two correct; 1 mark for one correct]

3. a) $\frac{15}{28}$ = 0.5357… [1]
   = 53.57…%
   = 54% (to the nearest whole number) [1]
   b) $\frac{4}{28}$ = 0.142857… = 14.2857…% [1]
   = 14% (to the nearest whole number) [1]

4. Percentage change = $\frac{\text{actual change}}{\text{original amount}}$ × 100%

   % change = $\frac{600}{5000}$ × 100% = 12% [1]
5. % change = $\frac{0.15}{1.4}$ × 100% = 10.714…% [1] = 10.7% (1 d.p.) [1]

1. $\frac{2}{5}$ = 40% [1]
   100% − 25% − 40% = 35% of the chickens are black [1]
2.   × 0.8
   Original price ⟶ New price
                    £99.30
               ÷ 0.8
   Original price = £99.20 ÷ 0.8 [1] = £124 [1]
3. 2430 × 1.2 [1] = £2916 [1]
   [1 mark for 1.2 or 0.2 × 2430 or 486]
4.   × 1.15
   Original price ⟶ New price
                    £74.52
               ÷ 1.15
   Original price = £74.52 ÷ 1.15 [1] = £64.80 [1]

5. a)   × 1.08
   Original price ⟶ New price
                     £240 000
                ÷ 1.08
   Original price = £240 000 ÷ 1.08 [1] = £222 222 [1]
   b) Percentage change = $\frac{\text{actual change}}{\text{original amount}}$ × 100%
   = $\frac{45 000}{195 000}$ × 100% = 23.076…% [1] = 23.1% (1 d.p.) [1]
6. Calculation to show that the original mass is 6.6666… kg [1]
             × 0.9
   Original mass ⟶ New mass
                    6 kg
             ÷ 0.9
   Original mass = 6 kg ÷ 0.9 = 6.6666… kg

Pages 18–22: Sequences

1. a) $\frac{1}{2}$ × 6 + 2 = 3 + 2 = 5 [1]
   b) −3 × 6 + 2 = −18 + 2 = −16 [1]
   c) (2 − 2) ÷ 6 = 0 ÷ 6 = 0 [1]
   d) (−46 − 2) ÷ 6 = −48 ÷ 6 = −8 [1]
2. a) Term-to-term rule is + 4 [1]
   b) Term-to-term rule is × 5 [1]
3. a) Adding 1 more each time, + 4, + 5, + 6, + 7, … [1]
   b) Fibonacci-type sequence, i.e. each number is the sum of the two terms that precede it [1]
4. A, C and D [2]
   [1 mark for two correct and none incorrect]
5. a) 5 × 10 + 3 = 53 [1]
   b) 88 [1] 103 [1]

> All terms in this sequence have units digits 3 or 8.

6. a) Input = 1: (1 + 20) × −2 = 21 × −2 = −42
   Input = 2: (2 + 20) × −2 = 22 × −2 = −44 [1]
   Input = 3: (3 + 20) × −2 = 23 × −2 = −46
   Input = 4: (4 + 20) × −2 = 24 × −2 = −48 [1]
   b) (−60 ÷ −2) − 20 or 30 − 20 [1]
   = 10 (10th term) [1]
7. a) 7, 10, 13, 34, 154, 304 [3]
   [2 marks for four correct terms; 1 mark for two correct terms]
   b) 5, 4, 3, −4, −44, −94 [3]
   [2 marks for four correct terms; 1 mark for two correct terms]
   c) −7.5, −7, −6.5, −3, 17, 42 [3]
   [2 marks for four correct terms; 1 mark for two correct terms]
   d) 13, 11, 9, −5, −85, −185 [3]
   [2 marks for four correct terms; 1 mark for two correct terms]

1. a) Increasing [1]
   b) Increasing [1]
   c) Decreasing [1]
   d) Decreasing [1]
2. a) Common difference is 3 [1]
   So a = 23 and b = 26 [1]
   b) Common difference is (31 − 19) ÷ 2 = 12 ÷ 2 = 6 [1]
   So c = 25 and d = 37 [1]
   c) Common difference is (4 − 10) ÷ 3 = −6 ÷ 3 = −2 [1]
   So e = 8 and f = 6 [1]
   d) Common difference is (15 − 9) ÷ 2 = 6 ÷ 2 = 3 [1]
   So g = 6 and h = 12 [1]
3. a) 4n [2]
   [1 mark for stating common difference is 4]
   b) 4n + 1 [2]
   [1 mark for 4n or for stating common difference is 4]
   c) 3n + 15 [2]
   [1 mark for 3n or for stating common difference is 3]
   d) 22 − 2n [2]
   [1 mark for −2n or for stating common difference is −2]
   e) 0.5n + 2.5 [2]
   [1 mark for 0.5n or for stating common difference is 0.5]
   f) 11 − n [2]
   [1 mark for −n or for stating common difference is −1]

4. a) Terms of first sequence are 8, 13, 18, 23, ... **[1]**
   Terms of second sequence are 2, 9, 16, 23, ... **[1]**
   b) 58 **[2]**
   [1 mark for correctly continuing either sequence for at least eight terms]
5. a) £11 **[1]**
   b) £$(n + 1)$ **[1]**
   c) $A = \frac{10 \times 13}{2}$ **[1]**
   $= £65$ **[1]**
   d) $A = \frac{19 \times 22}{2}$ or 19×11 **[1]**
   $= £209$ **[1]**

1. All square numbers with units digits 4 or 9 or numbers 4, 9, 49 and 64 **[2]**
   [1 mark for at least two correct numbers]
2. a)

Pattern number	1	2	3	4
Squares	2	3	4	5
Circles	1	3	6	10

[2]
   [1 mark for each row]
   b) $n + 1$ **[1]**
   c) Triangular numbers **[1]**
   d) $T = (10 \times 11) \div 2$ **[1]**
   $= 55$ **[1]**
   e) $T = \frac{n(n+1)}{2} + n + 1$ **[1]**
3. a) (25, 125) **[2]**
   [1 mark for each coordinate]
   b) (n^2, n^3) **[2]**
   [1 mark for each coordinate]

Pages 23–27: Area and Volume

1. a) It does not have a uniform cross-section. **[1]**
   b) E.g. cube, cuboid, triangular prism, pentagonal prism, hexagonal prism, cylinder. **[1 mark for any correct answer]**
2. $A = \frac{8 \times 4}{2} = 16$ **[1]**
   $V = 16 \times 20 = 320$ cm³ **[1]**
3. Volume C $= 6 \times 7 \times 5$ **[1]** $\times 2 = 210$ cm³ **[1]**
   Volume P $= A \times 10 = 210$ cm³ **[1]**
   Area $A = 21$ cm² **[1]**
4. Q (48 m²) **[1]** is larger by 12 m² **[1]** than P (36 m²) **[1]**
5. Area A $= (8 \times 9 + 8 \times 3 + 9 \times 3)$ **[1]** $\times 2 = 246$ cm² **[1]**
   Area A $= (6 \times 5 + 5 \times 10 + 6 \times 10)$ **[1]** $\times 2 = 280$ cm² **[1]**
   B has the greater area by 34 cm² **[1]**

   A cuboid has six faces. Work out the area of each face.

6. a) Two triangles **[1]** and three rectangles sketched **[1]**
   b) Area $= 6$ **[1]** $+ 6 + 24 + 32 + 40$ **[1]** $= 108$ cm² **[1]**

1. Area $= (8 \times 4)$ **[1]** $+ (6 \times 28) = 200$ cm² **[1]**
   Volume $= 200 \times 9$ **[1]** $= 1800$ cm³ **[1]**
2. Surface area $= 60 + 60 + 50 + 50 + 30 + 30$ **[2]** $= 280$ cm² **[1]**
   Cost $= 280 \div 40 = £7$ **[1]**
3. Lawn area $= 96$ **[1]** $- 8$ **[1]** $= 88$ m² **[1]**
   Cost $= 88 \times £6.75 = £594$ **[1]**
4. 57 m² **[1]** $- 3$ m² **[1]** $= 54$ m² **[1]**
5. Area $= \frac{6 \times 8}{2} = 24$ **[1]**
   Volume $= 24l = 2160$ cm³ **[1]**
   $l = \frac{2160}{24} = 90$ cm **[1]**
   Area of two triangles $= 24 \times 2 = 48$ **[1]**
   Area of three triangles $= (10 \times 90) + (6 \times 90) + (8 \times 90) = 2160$ **[1]**
   Surface area $= 48 + 2160 = 2208$ cm² **[1]**
6. $\frac{1}{2} \times \left(\frac{1}{10} + \frac{3}{20}\right) \times \frac{1}{8} = \frac{1}{2} \times \frac{1}{4} \times \frac{1}{8}$ **[1]** $= \frac{1}{64}$ m² **[1]**

1. Cross-sectional area $= 36 + 48$ **[1]** $= 84$ cm² **[1]**
   $84 + 84 + 60 + 60 + 200 + 200 + 240$ **[1]**
   Surface area $= 928$ cm² **[1]**

2. Area of one face $= 96 \div 6 = 16$ cm² **[1]**
   Side length $= 4$ cm **[1]**
   Volume $= 4 \times 4 \times 4$ **[1]** $= 64$ cm³ **[1]**
3. Total area $= 24n^2 + 12n^2 = 36n^2$ **[2]**
   [1 mark for $24n^2$ or $12n^2$]
   Side length $= 6n$ **[1]**

   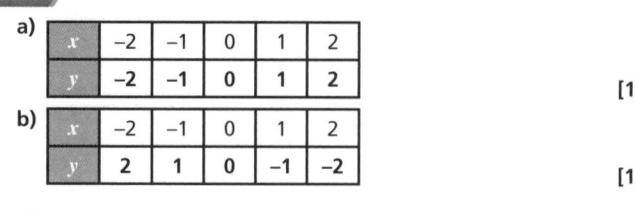
   Area of square $= 6n \times 6n = 36n^2$

4. Area of square $= 12 \times 12 = 144$ m² **[1]**
   Area of white triangles $= \frac{1}{2} \times 4 \times 7 + \frac{1}{2} \times 5 \times 12 + \frac{1}{2} \times 8 \times 12$ **[1]**
   Area of white triangles $= 14 + 30 + 48 = 92$ m² **[1]**
   Decking area $= 144 - 92 = 52$ m² **[1]**

   Cost $= 52 \div 4 \times 209.80$ **[1]** $= £2727.40$ **[1]**

Pages 28–32: Graphs

1. a)

x	−2	−1	0	1	2
y	−2	−1	0	1	2

[1]
   b)

x	−2	−1	0	1	2
y	2	1	0	−1	−2

[1]
   c)
   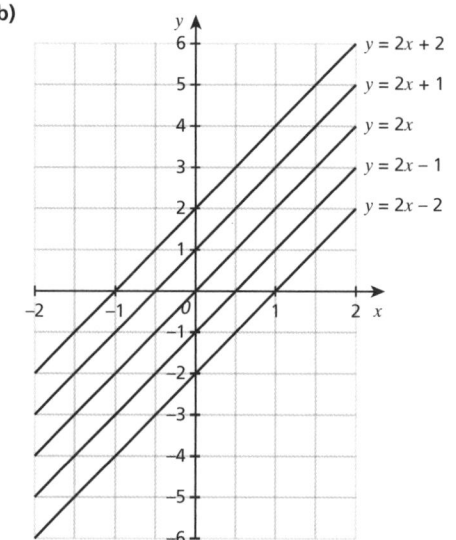
   [2]
   [1 mark for each line correctly drawn and labelled]
2. A (0, 1) **[1]**   B (0, −2) **[1]**   C (0, −1) **[1]**
3. a)

x	−2	−1	0	1	2
$y = 2x$	−4	−2	0	2	4
$y = 2x + 2$	−2	0	2	4	6
$y = 2x + 1$	−3	−1	1	3	5
$y = 2x − 1$	−5	−3	−1	1	3
$y = 2x − 2$	−6	−4	−2	0	2

[5]
   [1 mark for each correct row]
   b)
   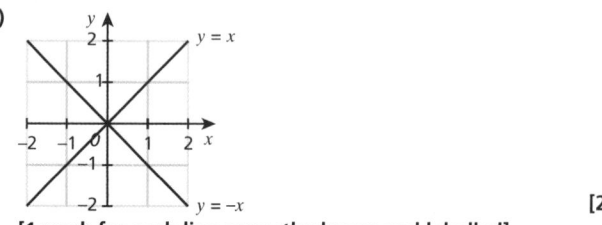
   [5]
   [1 mark for each correct line]
   c) **Any suitable answer, e.g.**
   The lines are parallel.
   The lines have the same gradient.
   The gradient is always 2.
   The y-intercept is always the same as the constant term **[1]**
   d) Any equation of the form $y = 2x + c$, e.g. $y = 2x + 3$ **[1]**

4. a) Given in question
   b) Given in question
   c) Gradient $= \frac{1}{3}$ [1]
   d) Gradient $= -\frac{2}{2} = -1$ [1]
   e) Gradient $= -\frac{4}{1} = -4$ [1]
   f) Gradient $= \frac{3}{2}$ [1]

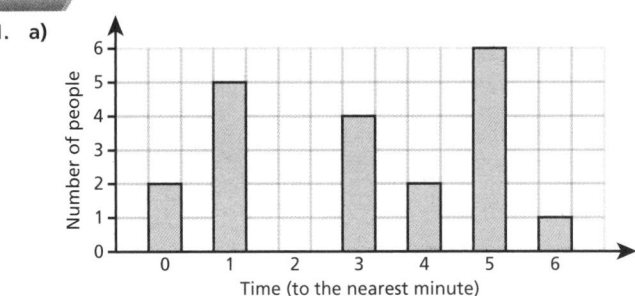

1. A Gradient $= 1$ [1]   **B** Gradient $= 2$ [1]
   C Gradient $= 3$ [1]   **D** Gradient $= \frac{1}{2}$ [1]
   E Gradient $= \frac{1}{3}$ [1]

2. a) Gradient $= 4$ y-intercept is (0, 1) [1]
   b) Gradient $= 4$ y-intercept is (0, –1) [1]
   c) Gradient $= 5$ y-intercept is (0, –3) [1]
   d) Gradient $= \frac{1}{2}$ y-intercept is (0, 2) [1]
   e) Rearranged equation: $y = -3x + 4$ [1]
      Gradient $= -3$ y-intercept is (0, 4) [1]
   f) Rearranged equation: $y = -\frac{1}{2}x - \frac{1}{2}$ [1]
      Gradient $= -\frac{1}{2}$ y-intercept is (0, $-\frac{1}{2}$) [1]

3. a) $y = 3x + 1$ [1] **b)** $y = 3x - 2$ [1] **c)** $y = 4x - 3$ [1]
   d) $y = -2x + 6$ [1] **e)** $y = \frac{1}{3}x + 4$ [1] **f)** $y = -\frac{1}{3}x - 2$ [1]

4. a) A $y = x$ [1] **B** $y = x + 1$ [1] **C** $y = x + 2$ [1]
   b) A $y = 2x$ [1] **B** $y = 2x + 2$ [1] **C** $y = 2x + 4$ [1]
   c) A $y = \frac{1}{2}x$ [1] **B** $y = \frac{1}{2}x + 1$ [1] **C** $y = \frac{1}{2}x + 2$ [1]

1. A $y = 3$ **D** $y = -2x - 2$
   B $y = 2x - 4$ **E** $y = -x$
   C $y = x + 2$ [3]
   [2 marks for three correct; 1 mark for one or two correct]

2. a)

Lines with gradient 3	Lines with gradient –3	Lines passing through (0, 3)	Lines passing through (0, –3)
$y = 3x + 3$	$y = -3x + 2$	$y = 3x + 3$	$y = 2x - 3$
$y = 3x - 4$	$y = -3x - 4$	$y = -3x + 3$	$y = 3x - 3$
$y = 3x - 3$	$y = -3x + 3$		$y = -3x - 3$
	$y = -3x - 3$		

[4]

[1 mark for each correct column]
   b) $y = 3x + 7$ [1]

Pages 33–35: Simplifying Numbers

1. a) 7.5 [1] **b)** 8.03 [1] **c)** 15.383 [1] **d)** 0.03 [1]
   e) 27.3 [1] **f)** 6.50 [1]

> Look at the next digit after the decimal place you want. If it is 5 or more, round up. If the question asks for 2 decimal places, you must write 2 decimal places, even if the second one is zero.

2. a) 60 [1] **b)** 100 [1] **c)** 5000 [1] **d)** 60 000 [1]

> The first significant figure is the one with the highest place value.
> To round to 1 significant figure, round to the place value of the first digit from the left.
> For example, in 546 the first significant figure has value 500, so round to the nearest 100.

3. a) 360 [1] **b)** 4100 [1] **c)** 56 000 [1] **d)** 21 000 [1]
   e) 350 000 [1] **f)** 210 000 [1] **g)** 1 500 000 [1] **h)** 13 000 000 [1]
4. a) 6720 [1] **b)** 98 500 [1] **c)** 3160 [1] **d)** 409 000 [1]
   e) 35 500 [1] **f)** 106 000 [1] **g)** 2 410 000 [1] **h)** 1 370 000 [1]
5. a) Estimate $20 \times 10 = 200$ [1]
   b) Estimate $400 \times 20 = 8000$ [1]
   c) Estimate $400 \times 30 = 12 000$ [1]
   d) Estimate $6000 \times 30 = 180 000$ [1]

1. a) 2 [1] **b)** 3 [1] **c)** 20 [1] **d)** 0.3 [1] **e)** 0.03 [1] **f)** 0.006 [1]
   g) 0.0002 [1] **h)** 0.05 [1]

2.

Number	1 s.f.	2 s.f.	3 s.f.
147	100	150	147
2095	2000	2100	2100
18.36	20	18	18.4
8079	8000	8100	8080
0.0351	0.04	0.035	0.0351
4872	5000	4900	4870
0.7599	0.8	0.76	0.760

3. a) 3550 [1] **b)** 3547.9 [1] **c)** 4000 [1]
   d) 3500 [1] **e)** 3547.86 [1] **f)** 3547.9 [1]
4. a) Estimate $\frac{5 \times 20}{10} = 5 \times 2 = 10$ [1]
   b) Estimate $\frac{90 + 30}{40} = \frac{120}{40} = 3$ [1]
   c) Estimate $\frac{7}{0.5} = 14$ [1]
   d) Estimate $0.2 \times 0.2 = 0.04$ [1]

1. a) 3 significant figures [1]
   b) Josh 19 542 rounds to 20 000 to 2 significant figures.
      Lee 19 542 rounds to 20 000 to 1 significant figure.
      Their answers are the same. [2]
      [1 mark if both are rounded to 20 000 but no comment made.]
2. a) Estimate $(30)^2 - 300 = 900 - 300 = 600$ [1]
   b) Estimate $\frac{0.4 \times 8}{0.4} = 8$ [1]
3. Area $=$ length $\times$ width $= 177 \times 540$
   Estimate $= 200 \times 500 = 100 000 \, \text{m}^2$ [1]
4. a) Volume of cube $= (4.8)^3$
      Estimate $5^3 = 125 \, \text{cm}^3$ [1]
   b) This is an overestimate. [1]
5. 18 450 g $\div$ 225 g [1]
   Estimate 20 000 $\div$ 200 $= 100$ [1]
   Or 18.45 kg $\div$ 0.225 kg [1]
   Estimate 20 $\div$ 0.2 $= 100$ [1]

Pages 36–41: Presenting and Interpreting Data

1. a)

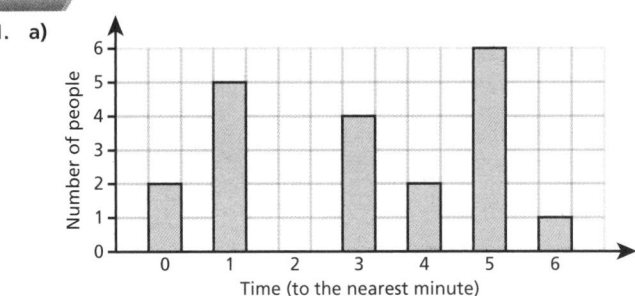

[2]

[1 mark for at least two more correct bars]
   b) Mean waiting time $=$
      $((2 \times 0) + (5 \times 1) + (0 \times 2) + (4 \times 3) + (2 \times 4) + (6 \times 5) + (1 \times 6))$
      $\div 20$
      or $(0 + 5 + 0 + 12 + 8 + 30 + 6) \div 20$ [1]
      $= 61 \div 20$ [1] $= 3.05$ or 3 minutes [1]

2. Child: $\frac{60}{360} \times 180$ or $\frac{1}{6} \times 180 = 30$ [1]
   Adult: $\frac{90}{360} \times 180$ or $\frac{1}{4} \times 180 = 45$ [1]
   Day pass: $\frac{120}{360} \times 180$ or $\frac{1}{3} \times 180 = 60$ [1]
   Weekly pass: $\frac{90}{360} \times 180$ or $\frac{1}{4} \times 180 = 45$ [1]

3. a) $\frac{1}{2} \times 60 = 30$ [1]

   b)

Bus	□ □ □	
Walk	□ □	Key:
Cycle	□	□ represents 10 students
Car	□	

   [3]

   [2 marks for two or three correct rows; 1 mark for one correct row]

4. a) Dog 50% **[1]**; Horse 20% **[1]**; Llama 20% **[1]**; Cat 10% **[1]**
b) 100 − 40 − 25 − 5 **[1]** = 30% **[1]**
c) i) Not definitely true as data shows proportions not actual numbers [1]
ii) Definitely true as dog most popular in both year groups **[1]**
iii) Definitely true as 25% chose cat in Year 8 and 10% in Year 7 [1]

1. a) Red: $\frac{6}{24} \times 360° = 90°$

Blue: $\frac{12}{24} \times 360° = 180°$

Yellow: $\frac{5}{24} \times 360° = 75°$

Green: $\frac{1}{24} \times 360° = 15°$

Number of students	24	1	5	6	12
Angle size	360°	15°	75°	90°	180°

Fully correct [2]
[1 mark for any correct calculation]
b)

Favourite colour

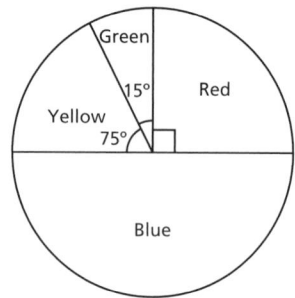

Fully correct and labelled [2]
[1 mark for correct sectors without labels]
2. a) Fewer walk or cycle [1]
b) Year 8 and Year 9 (26 fewer students walk or cycle, whereas 15 fewer between Year 7 and Year 8) [1]
c) Example answer: Difference is 15 and then 26 so if the trend continues the approximate difference will be 37 **[1]**, meaning that approximately 60–65 walk or cycle **[1]**
3. a) 360° ÷ 45° = 8 [1]
8 × 15 = 120 [1]
b) Each student is represented by 3° [1]
100° ÷ 3° is not an integer, so not possible [1]
4. a) The longer the ride, the more time taken [1]
b) As amount of rainfall decreased, the distance of ride increased **or** Rides tended to be shorter when raining [1]
c) The number of riders did not affect the distance of the ride **or** The distance of the ride was not affected by the number of riders [1]

1. a)

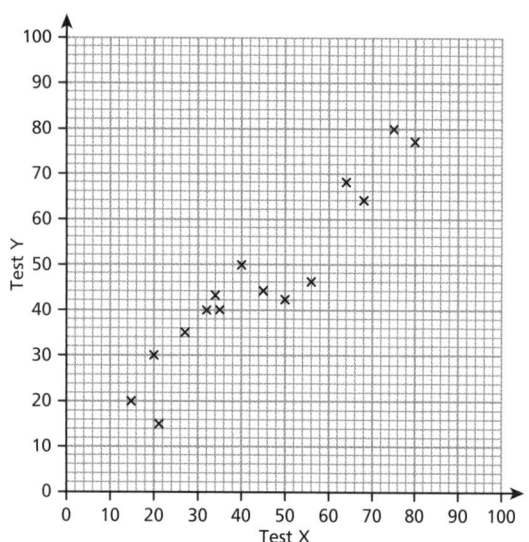

All points plotted correctly [4]

[3 marks for at least ten points plotted correctly; 2 marks for at least six points plotted correctly; 1 mark for at least three points plotted correctly]
b) Any suitable answer, e.g.
In general, the better students do on Test X, the better they do on Test Y. [1]
or Students who did well on Test X also did well on Test Y. **[1]**
2. a) 50 × 4 = 200 [1]
200 − (54 + 38 + 50) = 58 [1]
b) 85 + 92 = 177
Yes, as only need 23 in Round 3 to qualify [1]
c) 200 − 41 = 159 [1]
159 ÷ 3 = 53 [1]
3. a) $\frac{240}{360} \times 210$ or $\frac{2}{3} \times 210$ [1]
= 140 students [1]
b) Unlikely to be true [1]
Cannot assume same proportions for girls and boys [1]

Pages 42–45: Algebra

1. a) 3x + 12 [1]
b) 5x − 35 [1]
c) 14y + 28 [1]
d) 36y − 20 [1]
e) $a^2 + 3a$ [1]
f) $b^2 − 6b$ [1]
g) $2c^2 + 4c$ [1]
h) $6d^2 − 15d$ [1]

To expand a bracket, multiply every term in the bracket by the term on the outside.
$2(x + 5) = 2x + 10$

2. a) 3x + 4y [1]
b) 3r − s [1]
c) 12x − 5 [1]
d) 5t + 6v − 11 [1]

To simplify, add like terms:
• Add x terms to x terms.
• Add y terms to y terms.
• Add number terms to number terms.

3. Boxes joined as follows:
5 more than x   to   x + 5
5 less than x   to   x − 5
5 lots of x   to   5x
x divided by 5   to $\frac{x}{5}$ [2]
[1 mark for two correct]

4. a) 4 − 1 = 3 [1]
b) 4 − 1 + 2 = 5 [1]
c) 4 × 2 = 8 [1]
d) 4 × −1 = −4 [1]
e) $\frac{4}{2} = 2$ [1]
f) $\frac{4−1}{2} = 1.5$ [1]
g) 2(4 + 1) = 10 [1]
h) $(2 \times 4 + 3 \times −1)^2 = (8 − 3)^2 = 25$ [1]
5. a) t^4 **[1]**   **b)** $w^{2+5} = w^7$ **[1]**   **c)** x^2 **[1]**   **d)** $a^{2−1} = a$ **[1]**
e) b^4 **[1]**   **f)** $5c^2$ **[1]**   **g)** y **[1]**   **h)** $5z^5$ **[1]**

To multiply powers of the same letter, add the powers.
To divide powers of the same letter, subtract the powers.

6. a) 7n **[1]**   **b)** C = 7n **[1]**   **c)** $P = \frac{7n}{100}$ or P = 0.07n **[1]**
7. E = 18n [1]
8. $s = \frac{d}{t} = \frac{25}{3.5} = 7.142....$ **[1]** = 7.1 km/h (to 1 d.p.) **[1]**

1. a) −3x − 18 [1]
b) −2x + 6 [1]

c) $-x - 8$ [1]

d) $-2x^2 + 8x$ [1]

2. **a)** $3x + 6 + 4x - 4$ [1]
  $= 7x + 2$ [1]
  b) $5y - 35 - 2y - 6$ [1]
  $= 3y - 41$ [1]
  c) $8z - 16 - 3z + 12$ [1]
  $= 5z - 4$ [1]
  d) $x^2 + 3x - 2x - 2$ [1]
  $= x^2 + x - 2$ [1]
  e) $y^2 - 2y + 3y^2 + 12y$ [1]
  $= 4y^2 + 10y$ [1]
  f) $15z^2 + 3z - 4z^2 - 7z$ [1]
  $= 11z^2 - 4z$ [1]

Expand the brackets. Then simplify by adding like terms. x terms and x^2 terms are not like terms.

3. **a)** $5x^2$ [1] **b)** $3x^4$ [1] **c)** $6x^2$ [1] **d)** $4x^3$ [1]

Divide the numbers, then divide the letters.

4. **a)** length × width $= 2x^2$ [1]
  b) $5 \times 2x^2 = 10x^2$ [1]

5. Length $= w + 3$ [1]
  Area $= w(w + 3)$ or $w^2 + 3w$ [1]

6. **a)** $4(x + 2)$ [1]
  b) $5(x - 4)$ [1]
  c) $5(2x + 3)$ [1]
  d) $7(3x - 2)$ [1]
  e) $x(x - 1)$ [1]
  f) $x(3x + 1)$ [1]
  g) $2x(3x - 2)$ [1]
  h) $3x(4x + 5)$ [1]

7. Perimeter $= x + 4x - 3 + 2x + 1$ [1] $= 7x - 2$ [1]

1. **a)** $y = (7 - 2)^2 + 4 = 5^2 + 4 = 29$ [1]
  b) $y = (-2 - 2)^2 + 11 = (-4)^2 + 11 = 27$ [1]

2. Area of square $= 3x \times 3x = 9x^2$ [1]
  Area of rectangle $= x(x + 5) = x^2 + 5x$ [1]
  Shaded area $= 9x^2 - (x^2 + 5x) = 8x^2 - 5x$ or $x(8x - 5)$ [1]

3. Examples: $(x^3)^4 = x^3 \times x^3 \times x^3 \times x^3 = x^{12}$
  $(4x)^2 = 4x \times 4x = 16x^2$

  a) x^4 [1] **b)** y^6 [1] **c)** t^8 [1] **d)** $4n^2$ [1] **e)** $25r^6$ [1] **f)** $27w^6$ [1]

4. **a)** $65 = \frac{d}{3}$ [1]
  $d = 3 \times 65 = 195$ km [1]
  b) $18 = \frac{45}{t}$ [1]
  $18t = 45$
  $t = \frac{45}{18} = \frac{5}{2} = 2.5$ hours [1]

5. **a)** 75 [1] **b)** 49 [1] **c)** 12 [1] **d)** 10 [1] **e)** 4 [1]
  f) 7 [1] **g)** 30 [1] **h)** −7.5 [1]

6. Area $= \frac{1}{2}bh = \frac{1}{2} \times 2x(x^2 + 3)$ [1]
  $= x(x^2 + 3)$ or $x^3 + 3x$ [1]

7. $n = 1.1p$ [1]

Pages 46–48: Fractions and Decimals

1. **a)** 0.7 [1] **b)** 0.25 [1] **c)** 0.8 [1] **d)** 0.45 [1]
  e) 0.275 [1] **f)** 0.375 [1] **g)** 0.38 [1] **h)** 0.065 [1]

2. **a)** $\frac{3}{10}$ [1]
  b) $\frac{9}{100}$ [1]
  c) $\frac{14}{25}$ [1]
  d) $\frac{1}{8}$ [1]

3. **a)** $\frac{1}{40}$ [1]
  b) $\frac{8}{40} + \frac{5}{40} = \frac{13}{40}$ [1]

c) $\frac{1}{10} - \frac{1}{12} = \frac{6}{60} - \frac{5}{60} = \frac{1}{60}$ [1]

d) $\frac{1}{3} \div \frac{1}{4} = \frac{1}{3} \times 4 = 1\frac{1}{3}$ [1]

e) $\frac{10}{42} = \frac{5}{21}$ [1]

f) $\frac{35}{45} - \frac{18}{45} = \frac{17}{45}$ [1]

g) $\frac{3}{5} \times \frac{7}{3} = \frac{7}{5} = 1\frac{2}{5}$ [1]

h) $\frac{8}{36} + \frac{27}{36} = \frac{35}{36}$ [1]

Remember:
- **to multiply fractions**, multiply numerator by numerator and denominator by denominator; simplify if possible.
- **to divide by a fraction**, multiply by its reciprocal (e.g. the reciprocal of $\frac{3}{10}$ is $\frac{10}{3}$)
- **to add or subtract fractions**, convert one or both fractions so they have the same denominator before adding or subtracting.

4. $\frac{3}{8} = 0.375$, $\frac{5}{12} = 0.41666...$, $\frac{17}{20} = 0.85$
  20% $\frac{3}{8}$ 0.39 $\frac{5}{12}$ $\frac{17}{20}$ 0.876 [2]
  [1 mark if one value is incorrectly ordered]

5. **a)** $0.3 \times 4 = 3 \div 10 \times 4 = 3 \times 4 \div 10 = 1.2$ [1]
  b) 0.01 [1]
  c) 0.048 [1]
  d) 1.4 [1]

6. **a)** $1\frac{6}{8} + \frac{5}{8} = \frac{14}{8} + \frac{5}{8} = \frac{19}{8} = 2\frac{3}{8}$ [1]
  b) $\frac{9}{4} - \frac{3}{4} = \frac{6}{4} = 1\frac{1}{2}$ [1]
  c) $\frac{17}{12} = 1\frac{5}{12}$ [1]
  d) $8\frac{5}{6}$ [1]
  e) $6\frac{1}{10}$ [1]
  f) $6\frac{11}{35}$ [1]

7. **a)** $1\frac{2}{5} \times 2\frac{1}{4} = \frac{7}{5} \times \frac{9}{4} = \frac{63}{20} = 3\frac{3}{20}$ [1]
  b) $4\frac{5}{8} \div 1\frac{1}{2} = \frac{37}{8} \div \frac{3}{2} = \frac{37}{8} \times \frac{2}{3} = \frac{74}{24} = 3\frac{2}{24} = 3\frac{1}{12}$ [1]
  c) $2\frac{11}{32}$ [1]
  d) $29\frac{7}{10}$ [1]

To multiply or divide mixed numbers, first convert them to improper fractions.

1. **a)** 0.333 (3 d.p.) [1]
  b) 0.429 (3 d.p.) [1]
  c) 0.444 (3 d.p.) [1]
  d) 0.778 (3 d.p.) [1]

2. **a)** $5\frac{1}{2} \times 3\frac{2}{3}$ [1] $= \frac{121}{6} = 20\frac{1}{6}$ inches² [1]
  b) $\frac{1}{2} \times 6\frac{1}{4} \times 15$ [1] $= \frac{1}{2} \times \frac{25}{4} \times 15 = \frac{375}{8} = 46\frac{7}{8}$ inches² [1]

3. **a)** Term-to-term rule is ÷ 5
  Next three terms are $\frac{4}{5}, \frac{4}{25}, \frac{4}{125}$ [3]
  [1 mark for each correct term]
  b) Term-to-term rule is × $1\frac{1}{2}$
  Next three terms are: $10\frac{1}{8}, 15\frac{3}{16}, 22\frac{25}{32}$ [3]
  [1 mark for each correct term]

4. Measure the angles.

  a) Romance 50°
  $\frac{50}{360} = \frac{5}{36}$ [1]
  b) Thriller 140°
  $\frac{140}{360} = \frac{7}{18}$ [1]
  c) Action 70°, which is half the angle for Thriller.
  $\frac{7}{36}$ [1]

5. a) $\frac{1}{12}$ of $113\,832 = \frac{47\,430}{5}$
$= 9486$ [1]
b) $\frac{5}{6}$ of $113\,832 = 2 \times \frac{5}{12} \times 113\,832$
$= 2 \times 47\,430 = 94\,860$ [1]
6. $1\frac{2}{5}$ $1\frac{1}{3}$ 133% 1.2 102% [2]
[1 mark if one value is incorrectly ordered]

1. 6th term $= \frac{11}{12}$ [1]

> The first term less than 1 is the first term where the numerator is less than the denominator.

2. a) Term-to-term rule is $+ \frac{1}{4}$
Next three terms: $1\frac{3}{20}, 1\frac{2}{5}, 1\frac{13}{20}$ [3]
[1 mark for each correct term]
b) Term-to-term rule is $- \frac{1}{2}$
Next three terms: $2\frac{1}{18}, 1\frac{5}{9}, 1\frac{1}{18}$ [3]
[1 mark for each correct term]

3. Term-to-term rule is $+ \frac{1}{2}$
nth term $= \frac{1}{2}n + 4\frac{3}{4}$ [2]
[1 mark for $\frac{1}{2}n$]

4.
> Ascending order means from smallest to greatest.

$-\frac{3}{4}$ -70% -0.6 $\frac{3}{8}$ $\frac{2}{3}$ [2]
[1 mark if one is incorrectly ordered]

5.
> To find the value halfway between two numbers, work out their mean.

$1\frac{3}{5} + 2\frac{5}{8} = \frac{169}{40}$ [1]
Half of $\frac{169}{40} = \frac{169}{80} = 2\frac{9}{80}$ [1]

6.
> If $2\frac{1}{4}$ is in the sequence, this equation will have a solution where n is a whole number.

$\frac{n}{4} - 3 = 2\frac{1}{4}$ [1]
$\frac{n}{4} = 5\frac{1}{4} = \frac{21}{4}$
$n = 21$
Yes, $2\frac{1}{4}$ is in the sequence, it is the 21st term. [1]

Pages 49–51: Proportion

1. a) £8.05 ÷ 7 = £1.15 [1]
b) £3.45 [1]
2. a) 160 [1]
b) 16 [1]
3. a) 50g flour [1]   1 egg [1]   150ml milk [1]
$\frac{1}{2}$ teaspoon sunflower oil [1]
b) 250g flour [1]   5 eggs [1]   750ml milk [1]
$2\frac{1}{2}$ teaspoons sunflower oil [1]
4. a) 160km [1]
b) 40km [1]
c) 20km [1]
5. a) 6 hours [1]
b) 1 hour [1]
c) $1\frac{1}{2}$ hours [1]

1.

Litres	0	1	4	8	10
Pints	0	1.75	7	14	17.5

[5]

2. a) 20 minutes [1]
b) 1 machine would take 3 hours to fill 270 bottles [1]
$1\frac{1}{2}$ hours [1]

3.

Number of machines	1	2	4	8	10
Time to fill 1500 pies	2 hours	1 hour	30 minutes	15 minutes	12 minutes

[5]

4. a) 60 litres [1]
b) $\frac{1}{5}$ hour = 12 minutes [1]
c) 20 × 7 × 24 = 3360 litres [1]

1.

Kilometres	0	1	8	16	24	32
Miles	0	$\frac{5}{8}$ or 0.625	5	10	15	20

[6]

2. 1 ticket costs £31.25 ÷ 5 = £6.25 [1]
$C = 6.25n$ [1]
3. a) From the graph, it costs £280 to hire the car for 4 days. The hire cost and the cost are in direct proportion, because the graph is a straight line through (0, 0). So it costs £280 × 2 = £560 to hire the car for 8 days [1]
b) 1 day's hire costs £70
$C = 70d$ [1]
4. 8kg = 17.6 pounds, so 18 pounds is heavier than 8kg [1]
5. Graphs matched as follows:
First graph   to   Not in proportion
Second graph   to   Direct proportion
Third graph   to   Inverse proportion
[2 marks if all correct; 1 mark if one correct]

Pages 52–57: Circles and Cylinders

1. a) Circumference = 50.3cm [1]   Area = 201.1cm² [1]
b) Circumference = 31.4cm [1]   Area = 78.5cm² [1]

> Check if you are given the diameter or radius before you begin.

2. Circle 56.5cm [1] greater than square (56cm) by 0.5cm [1]
3. Circumference = 1570.8m [1] × 3 = 4712.4m [1] = 4.7km [1]

> 1km = 1000m

4. a) Area $= \pi r^2$ so $314 = 3.14 \times r^2$ or $r^2 = 100$ [1]
$r = 10$cm [1]
b) $C = \pi \times d$ so $157 = 3.14 \times d$ or $d = 157 \div 3.14$ [1]
$d = 50$ cm [1]
5. Rectangle = 63cm² [1]
Circle = 12.6cm² [1]
Shaded area = 50.4cm² [1]

1. $\frac{1}{4} \times \pi \times 6.8^2$ [1] = 36.3cm² [1]
2. Semicircle $= \frac{1}{2} \times \pi \times 6^2$ [1] = 56.5m² [1]
Total area = 56.5 + (7 × 12) = 140.5m² [1]
140.5 ÷ 4 = 35.125, so 36 bags needed [1]
Cost = 36 × 28.95 = £1042.20 [1]
3. Volume $= (\pi \times 3^2)$ [1] × 12 [1] = 108π cm³ [1]

> Area of circle $= \pi r^2$

4. a) $\pi \times 5^2$ [1] = 25π cm² [1]
b) Triangle $= \frac{1}{2} \times 6 \times 8 = 24$cm² [1]
Shaded area = $(25\pi - 24)$ cm² [1]

> You cannot subtract 24 from 25π because they are not like terms.

5. Semicircle $= \frac{1}{2} \times \pi \times 35 + 35$ [1] = 89.98cm [1]
Quarter circle $= \frac{1}{4} \times \pi \times 40 + 40$ [1] = 71.42cm [1]
Cost = (89.98 + 71.42) × 5 [1] = 807p = £8.07 [1]

> A common mistake is forgetting to add on the lengths of the straight lines when finding the perimeter.

6. a) $\pi \times 7^2 \times 15$ [1] = 2309.0706 = 2309.071cm³ (3 d.p.) [1]
b) 2309.0706 cm³ = 2309.0706 ml = 2.3 litres [1]

1. 2×31.4 **[1]** $= 62.8$ cm **[1]**
2. Large circle $= \pi \times 10^2$ **[1]** $= 100\pi$ **[1]**
   Small circle $= \pi \times 9^2 = 81\pi$ **[1]**
   Area of path $= 100\pi - 81\pi = 19\pi$ m^2 **[1]**
3. Volume = area $\times 5 = \pi r^2 \times 5 = 80\pi$ **[1]**
   $r^2 = 80\pi \div 5\pi = 16$ **[1]**
   Radius = 4 cm **[1]**
   New volume $= \pi \times 8^2 \times 5$ **[1]** $= 320\pi$ **[1]**
4. $C = \pi \times 7.2 = 22.6$ cm **[1]**
   $\frac{2010}{22.6}$ or $\frac{20.1}{0.226}$ **[1]** = 88.9, so 88 complete rotations **[1]**
5. a) $\pi \times 50 + 200$ **[1]** $= 357.1$ m **[1]**
   b) $\frac{357.1}{7.5}$ **[1]** $\times 2 = 95.2$ s **[1]**
6. Semicircle $= \frac{1}{2} \times \pi \times 9.5^2 = 141.8$ cm^2
   [2 marks: 1 for method; 1 for accuracy]
   Rectangle width = 9.5 cm **[1]**
   Shaded area $= (19 \times 9.5) - 141.8 = 38.7$ cm^2 **[1]**
7. Let h = height of water
   Volume $= \pi \times 0.65^2 \times h = 0.265$ **[1]**
   $1.3273h = 0.265h = 0.265$
   $1.3273 = 0.19964...$ m **[1]**
   Height = 19.96 cm = 20 cm (to nearest cm) **[1]**

Pages 58–63: Equations and Formulae

1. a) $a = 3$ **[1]** b) $b = 4$ **[1]** c) $c = -3$ **[1]**
   d) $d = 4$ **[1]** e) $e = 19$ **[1]** f) $f = 6$ **[1]**
2. a) $a = 6$ **[1]** b) $b = 7$ **[1]** c) $c = 10$ **[1]** d) $d = 0.5$ or $d = \frac{1}{2}$ **[1]**
   e) $e = -6$ **[1]** f) $f = -\frac{3}{4}$ **[1]** g) $g = 16$ **[1]** h) $h = 0$ **[1]** i) $x = -16$ **[1]**
3. $a + 2a + 3a = 180°$ or $6a = 180°$ **[1]**
   $a = 30°$
   $3a = 90°$ (or largest angle is 90°) **[1]**
4. a) $5a = 21 - 1$ or $5a = 20$ **[1]**
   $a = 4$ **[1]**
   b) $3b = 23 + 4$ or $3b = 27$ **[1]**
   $b = 9$ **[1]**
   c) $2c = 10 - 1$ or $2c = 9$ **[1]**
   $c = 4.5$ **[1]**
   d) $4d = 13 + 7$ or $4b = 20$ **[1]**
   $d = 5$ **[1]**
   e) $2e = 6 - 6$ or $2e = 0$ **[1]**
   $e = 0$ **[1]**
   f) $5f = -3 - 2$ or $5f = -5$ **[1]**
   $f = -1$ **[1]**
   g) $3g = 15 - 24$ or $3g = -9$ **[1]**
   $g = -3$ **[1]**
   h) $\frac{h}{2} = 15 - 1$ or $\frac{h}{2} = 14$ **[1]**
   $h = 28$ **[1]**
   i) $\frac{x}{3} = 4 + 2$ or $\frac{x}{3} = 6$ **[1]**
   $x = 18$ **[1]**
5. a) $2x$ cm **[1]**
   b) $2x + x + 2x + x = 36$ or $6x = 36$ **[1]**
   $x = 6$, Width = 6 cm **[1]**
6. a) $t = s - 10$ **[1]**
   b) $t = v + 4.5$ **[1]**
   c) $t = \frac{x}{5}$ **[1]**
   d) $t = 2y$ **[1]**
7. a) $y - 3 = 2x$ **[1]**
   $\frac{y-3}{2} = x$ or $x = \frac{y-3}{2}$ **[1]**
   b) $T - y = x + z$ or $T - z = x + y$ **[1]**
   $T - y - z = x$ or $x = T - y - z$ **[1]**
   c) $5x = p + 4$ **[1]**
   $x = \frac{p+4}{5}$ **[1]**
   d) $6x = 8 - w$ or $w - 8 = -6x$ **[1]**
   $x = \frac{8-w}{6}$ or $\frac{w-8}{-6} = x$ **[1]**

1. a) $4x + 20 = 24$ or $x + 5 = 6$ **[1]**
   $4x = 24 - 20$ or $4x = 4$ or $x = 6 - 5$ **[1]**
   $x = 1$ **[1]**
   b) $4x + 20 = 3x + 24$ **[1]**
   $4x - 3x = 24 - 20$ **[1]**
   $x = 4$ **[1]**
   c) $4x + 20 = 2x - 14$ **[1]**
   $4x - 2x = -14 - 20$ or $2x = -34$ **[1]**
   $x = -17$ **[1]**
2. a) $5x - x = 12$ or $4x = 12$ **[1]**
   $x = 3$ **[1]**
   b) $5x - x = 19 - 7$ or $4x = 12$ **[1]**
   $x = 3$ **[1]**
   c) $6x - 2x = 10 + 2$ or $4x = 12$ **[1]**
   $x = 3$ **[1]**
3. a) $3a + 6 = 15$ or $a + 2 = 5$ **[1]**
   $3a = 15 - 6$ or $3a = 9$ or $a = 5 - 2$ **[1]**
   $a = 3$ **[1]**
   b) $3b + 6 = b$ **[1]**
   $3b - b = -6$ or $2b = -6$ **[1]**
   $b = -3$ **[1]**
   c) $3c + 6 = c + 15$ **[1]**
   $3c - c = 15 - 6$ or $2c = 9$ **[1]**
   $c = 4.5$ **[1]**
   d) $d - 3 = 40$ or $\frac{1}{2}d - \frac{3}{2} = 20$ **[1]**
   $d = 40 + 3$ or $\frac{1}{2}d = 20 + \frac{3}{2}$ or $\frac{1}{2}d = \frac{43}{2}$ **[1]**
   $d = 43$ **[1]**
   e) $e - 3 = 2e$ or $\frac{1}{2}e - \frac{3}{2} = e$ **[1]**
   $-3 = 2e - e$ or $-\frac{3}{2} = e - \frac{1}{2}e$ or $-\frac{3}{2} = \frac{1}{2}e$ **[1]**
   $e = -3$ **[1]**
   f) $f - 3 = 2f + 40$ or $\frac{1}{2}f - \frac{3}{2} = f + 20$ **[1]**
   $-3 - 40 = 2f - f$ or $-\frac{3}{2} - 20 = f - \frac{1}{2}f$ or $-\frac{43}{2} = \frac{1}{2}f$ **[1]**
   $-43 = f$ **[1]**
4. a) $3(x - 2) = 5(x - 6)$ **[1]**
   $3x - 6 = 5x - 30$ **[1]**
   $-6 + 30 = 5x - 3x$ or $24 = 2x$ **[1]**
   $x = 12$ **[1]**
   b) Perimeter $= 3 \times 10$ or 5×6 **[1]**
   = 30 cm **[1]**
5. a) $y - 4 = x$ **[1]**
   b) $2y - 4 = x$ **[1]**
   c) $y - 4 = 2x$ **[1]**
   $\frac{y-4}{2} = x$ **[1]**
   d) $y + 4 = 2x$ **[1]**
   $\frac{y+4}{2} = x$ **[1]**
   e) $y - 3w = 2x$ **[1]**
   $\frac{y-3w}{2} = x$ **[1]**
6. $40 \div \frac{1}{2}(7 + 3)$ **[1]** = 8 cm **[1]**

1. a) $x + 4 = 12$ **[1]**
   $x = 12 - 4$, $x = 8$ **[1]**
   b) $x - 4 = 22$ **[1]**
   $x = 22 + 4$, $x = 26$ **[1]**
   c) $x - 4 = 35$ **[1]**
   $x = 35 + 4$, $x = 39$ **[1]**
2. $3(x + 5) = 8x$ **[1]**
   $3x + 15 = 8x$ **[1]**
   $15 = 8x - 3x$ or $15 = 5x$ **[1]**
   $x = 3$ **[1]**
   So length of line $= 8 \times 3$ or $6 \times 4 = 24$ **[1]**
   Alternative method:
   Assume length of first line = 24 cm, so $3(x + 5) = 24$ **[1]**
   $3x + 15 = 24$ or $x + 5 = 8$ **[1]**
   $3x = 24 - 15$ or $x = 8 - 5$ **[1]**
   $x = 3$ **[1]**
   Check for second line: Each part is $2 \times 3 = 6$ cm so $4 \times 6 = 24$ or $8 \times 3 = 24$ so both lines have length 24 cm. **[1]**
3. a) $x + 4 = 24$ **[1]**
   $x = 20$ **[1]**
   b) $2(x + 4) = 24$ or $2x + 8 = 24$ **[1]**
   $2x = 24 - 8$ or $2x = 16$ or $x + 4 = 12$ or $x = 12 - 4$ **[1]**
   $x = 8$ **[1]**

c) $2(x + 4) = 3x + 24$ or $2x + 8 = 3x + 24$ [1]
$8 - 24 = 3x - 2x$ or $-16 = 3x - 2x$ [1]
$x = -16$ [1]
4. $2(16 + 8x) = 128$ or $16 + 8x = 64$ **[2]** so $x = 6$ **[1]**
[1 mark for attempting to solve either equation]
5. 1.1 litres = 110 cl [1]
$6g + 10 = 2g + 110$ [1]
$6g - 2g = 110 - 10$ or $4g = 100$ [1]
$g = 25$ [1]
Small bottle holds 50 cl, large bottle holds 160 cl [1]
6. $2x + 8 = 3x + 3$ [1]
$x = 5$ [1]
Length of rectangle = $2 \times 5 + 8$ or $3 \times 5 + 3 = 18$ cm [1]
Area of rectangle = 18×2 [1]
= 36 cm² [1]
One side of square is $\sqrt{36} = 6$ cm [1]

Pages 64–69: Statistics: Averages and Range

1. a) $15 - 1 = 14$ [1]
   b) $8 - -5 = 13$ [1]
2. a) Mode = 6 [1]
   b) Writing scores in order 1, 4, 6, 6, 6, 8, 8, 9, 9, 10, 10 [1]
   Median = 8 [1]
   c) Mean = $(1 + 4 + 6 + 6 + 6 + 8 + 8 + 9 + 9 + 10 + 10) \div 11$ or
   $77 \div 11$ [1]
   = 7 [1]
   d) Range = $10 - 1 = 9$ [1]
3. a) 6°C [1]
   b) $12 - 5 = 7$°C [1]

1. a) Writing in order 0, 4, 4, 5, 6, 7, 8, 11 [1]
   Median = 5.5 [1]
   b) Median = $100 + 5.5 = 105.5$ [1]
2. a) Could be true or false [1]
   (you cannot assume that half the students in class interval take half the time)
   b) Definitely false [1]
   (would be included in the class $0 < t \leqslant 10$)
   c) Could be true or false [1]
   (e.g. shortest time could be 2 minutes with longest time 24 minutes, giving a range of 22 minutes)
   d) Definitely true [1]
   (as 6 students took more than 20 minutes)
3. Team A has higher mean higher so better scores [1]
   Team A has lower range so more consistent [1]
4. a) $(5 \times 0) + (7 \times 1) + (3 \times 2) + (4 \times 3) + (0 \times 4) + (1 \times 5)$
   or $0 + 7 + 6 + 12 + 0 + 5$ or 30 [1]
   $30 \div 20$ [1]
   Mean = 1.5 [1]
   b) More as scored more than the mean for 20 matches [1]

1. a) Range = $85 - 18$ [1]
   = 67 [1]
   b) Range = 67 [1]
2. a) Median is $x + 4$ **[1]** = 5
   So $x = 1$ [1]
   b) Mode is $x + 1 = 2$ [1]
3. a)

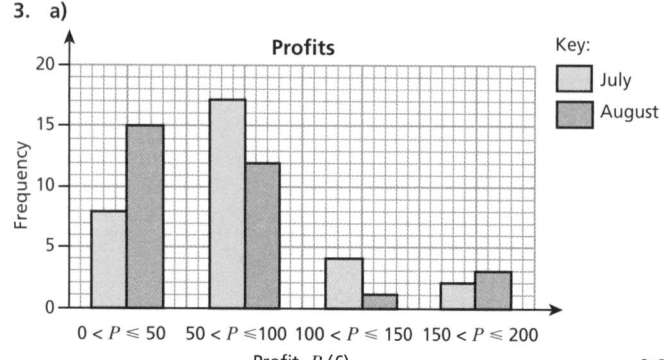

[3]
[1 mark for each correct pair of bars]

b) **Any suitable answers, e.g.** July more profitable, more days in August with profits less than £50 **or** July more profitable, more days in July with profits greater than £50. **[2]**
c) Not possible to compare the ranges as the lowest and highest profit each month is not known. **[1]**
4. a) Player A mean is $35 \div 5 = 7$; range is $8 - 6 = 2$ [1]
   Player B mean is $28 \div 5 = 5.6$; range is $10 - 2 = 8$ [1]
   Player C mean is $33 \div 5 = 6.6$; range is $7 - 6 = 1$ [1]
   Jo is player C, Imran is player A, Bob is player B [1]
   b) Third (Bob had lowest mean) [1]
5. a) **Worker A**

Job number	1	2	3	4	5	6
Time taken, t (minutes)	30	30	100	20	120	55

[2]
[1 mark for at least three correct cells]

Worker B

Job number	1	2	3	4	5	6	7
Time taken, t (minutes)	20	40	100	60	30	30	80

[2]
[1 mark for at least three correct cells]
b) Worker A: Range is $120 - 20 = 100$ minutes [1]
   Worker B: Range is $100 - 20 = 80$ minutes [1]
c) Worker A: Total time taken is $30 + 30 + 100 + 20 + 120 + 55$
   = 355 minutes [1]
   Worker B: Total time taken is $20 + 40 + 100 + 60 + 30 + 30 + 80 = 360$ minutes, so worker B [1]
d) **Worker A**

Time taken, t (minutes)	Frequency
$0 < t \leqslant 30$	3
$30 < t \leqslant 60$	1
$60 < t \leqslant 90$	0
$90 < t \leqslant 120$	2

[2]
[1 mark for two correct cells]

Worker B

Time taken, t (minutes)	Frequency
$0 < t \leqslant 30$	3
$30 < t \leqslant 60$	2
$60 < t \leqslant 90$	1
$90 < t \leqslant 120$	1

[2]
[1 mark for two correct cells]
e)

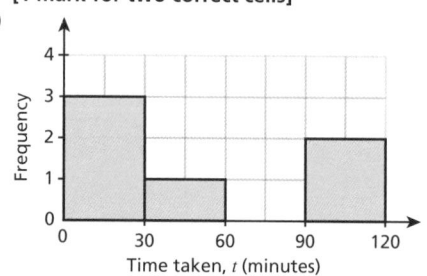

[2]
[1 mark for each correct bar]